AN ASSESSMENT OF THE PROSPECTS FOR
INERTIAL FUSION ENERGY

Committee on the Prospects for Inertial Confinement Fusion Energy Systems

Board on Physics and Astronomy

Board on Energy and Environmental Systems

Division on Engineering and Physical Sciences

NATIONAL RESEARCH COUNCIL
OF THE NATIONAL ACADEMIES

THE NATIONAL ACADEMIES PRESS
Washington, D.C.
www.nap.edu

THE NATIONAL ACADEMIES PRESS   500 Fifth Street, NW   Washington, DC 20001

NOTICE: The project that is the subject of this report was approved by the Governing Board of the National Research Council, whose members are drawn from the councils of the National Academy of Sciences, the National Academy of Engineering, and the Institute of Medicine. The members of the committee responsible for the report were chosen for their special competences and with regard for appropriate balance.

Support for this project was provided by Contract 10NA001274 between the National Academy of Sciences and the Department of Energy and the National Nuclear Security Administration. Any opinions, findings, conclusions, or recommendations expressed in this publication are those of the authors and do not necessarily reflect the views of the agencies that provided support for the project.

Cover: Ultraviolet laser beams aim at a fuel pellet before they implode it in the OMEGA laser target chamber. © Roger Ressmeyer/CORBIS

International Standard Book Number-13:   978-0-309-27081-6
International Standard Book Number-10:   0-309-27081-2

Copies of this report are available free of charge from:

Board on Physics and Astronomy
National Research Council
The Keck Center of the National Academies
500 Fifth Street, NW
Washington, DC 20001

Additional copies of this report are available from the National Academies Press, 500 Fifth Street, NW, Keck 360, Washington, DC 20001; (800) 624-6242 or (202) 334-3313; http://www.nap.edu.

Copyright 2013 by the National Academy of Sciences. All rights reserved.

Printed in the United States of America

THE NATIONAL ACADEMIES
Advisers to the Nation on Science, Engineering, and Medicine

The National Academy of Sciences is a private, nonprofit, self-perpetuating society of distinguished scholars engaged in scientific and engineering research, dedicated to the furtherance of science and technology and to their use for the general welfare. Upon the authority of the charter granted to it by the Congress in 1863, the Academy has a mandate that requires it to advise the federal government on scientific and technical matters. Dr. Ralph J. Cicerone is president of the National Academy of Sciences.

The National Academy of Engineering was established in 1964, under the charter of the National Academy of Sciences, as a parallel organization of outstanding engineers. It is autonomous in its administration and in the selection of its members, sharing with the National Academy of Sciences the responsibility for advising the federal government. The National Academy of Engineering also sponsors engineering programs aimed at meeting national needs, encourages education and research, and recognizes the superior achievements of engineers. Dr. Charles M. Vest is president of the National Academy of Engineering.

The Institute of Medicine was established in 1970 by the National Academy of Sciences to secure the services of eminent members of appropriate professions in the examination of policy matters pertaining to the health of the public. The Institute acts under the responsibility given to the National Academy of Sciences by its congressional charter to be an adviser to the federal government and, upon its own initiative, to identify issues of medical care, research, and education. Dr. Harvey V. Fineberg is president of the Institute of Medicine.

The National Research Council was organized by the National Academy of Sciences in 1916 to associate the broad community of science and technology with the Academy's purposes of furthering knowledge and advising the federal government. Functioning in accordance with general policies determined by the Academy, the Council has become the principal operating agency of both the National Academy of Sciences and the National Academy of Engineering in providing services to the government, the public, and the scientific and engineering communities. The Council is administered jointly by both Academies and the Institute of Medicine. Dr. Ralph J. Cicerone and Dr. Charles M. Vest are chair and vice chair, respectively, of the National Research Council.

www.national-academies.org

COMMITTEE ON THE PROSPECTS FOR INERTIAL CONFINEMENT FUSION ENERGY SYSTEMS

RONALD C. DAVIDSON, Princeton University, *Co-Chair*
GERALD L. KULCINSKI, University of Wisconsin-Madison, *Co-Chair*
CHARLES BAKER, University of California at San Diego (retired)
ROGER BANGERTER, E.O. Lawrence Berkeley National Laboratory (retired)
RICCARDO BETTI, University of Rochester
JAN BEYEA, Consulting in the Public Interest (CiPI)
ROBERT L. BYER, Stanford University
FRANKLIN CHANG-DIAZ, Ad Astra Rocket Company
STEVEN C. COWLEY, United Kingdom Atomic Energy Authority
RICHARD L. GARWIN, IBM Thomas J. Watson Research Center
DAVID A. HAMMER, Cornell University
JOSEPH S. HEZIR, EOP Group, Inc.
KATHRYN McCARTHY, Idaho National Laboratory
LAWRENCE T. PAPAY, PQR, LLC
KEN SCHULTZ, General Atomics (retired)
ANDREW M. SESSLER, E.O. Lawrence Berkeley National Laboratory
JOHN SHEFFIELD, University of Tennessee at Knoxville
THOMAS A. TOMBRELLO, JR., California Institute of Technology
DENNIS G. WHYTE, Massachusetts Institute of Technology
JONATHAN S. WURTELE, University of California at Berkeley
ROSA YANG, Electric Power Research Institute, Inc.

MALCOLM McGEOCH, Consultant, PLEX, LLC

Staff

DAVID LANG, Program Officer, Board on Physics and Astronomy, *Study Director*
GREG EYRING, Senior Program Officer, Division on Engineering and Physical Sciences
TERI THOROWGOOD, Administrative Coordinator, Board on Physics and Astronomy
JONATHAN YANGER, Senior Project Assistant, Board on Energy and Environmental Systems
ERIN BOYD, Christine Mirzayan Science and Technology Policy Graduate Fellow (January-April 2011)
SARAH NELSON-WILK, Christine Mirzayan Science and Technology Policy Graduate Fellow (January-April 2012)

JAMES LANCASTER, Director, Board on Physics and Astronomy
JAMES ZUCCHETTO, Director, Board on Energy and Environmental Systems

BOARD ON PHYSICS AND ASTRONOMY

PHILIP H. BUCKSBAUM, Stanford University, *Chair*
DEBRA M. ELMEGREEN, Vassar College, *Vice Chair*
RICCARDO BETTI, University of Rochester
ADAM S. BURROWS, Princeton University
TODD DITMIRE, University of Texas
NATHANIEL J. FISCH, Princeton University
PAUL FLEURY, Yale University
S. JAMES GATES, University of Maryland
LAURA H. GREENE, University of Illinois, Urbana-Champaign
MARTHA P. HAYNES, Cornell University
MARK B. KETCHEN, IBM Thomas J. Watson Research Center
MONICA OLVERA de la CRUZ, Northwestern University
PAUL SCHECHTER, Massachusetts Institute of Technology
BORIS SHRAIMAN, Kavli Institute of Theoretical Physics
MICHAEL S. TURNER, University of Chicago
ELLEN D. WILLIAMS, BP International
MICHAEL WITHERELL, University of California at Santa Barbara

Staff

JAMES LANCASTER, Director
DONALD C. SHAPERO, Senior Scholar
DAVID LANG, Program Officer
CARYN JOY KNUTSEN, Associate Program Officer
TERI THOROWGOOD, Administrative Coordinator
BETH DOLAN, Financial Associate

BOARD ON ENERGY AND ENVIRONMENTAL SYSTEMS

ANDREW BROWN, JR., Delphi Corporation, *Chair*
WILLIAM BANHOLZER, The Dow Chemical Company
MARILYN BROWN, Georgia Institute of Technology
WILLIAM CAVANAUGH, Progress Energy (retired), Raleigh, North Carolina
PAUL A. DeCOTIS, Long Island Power Authority
CHRISTINE EHLIG-ECONOMIDES, Texas A&M University, College Station
SHERRI GOODMAN, CNA, Alexandria, Virginia
NARAIN HINGORANI, Consultant, San Mateo, California
ROBERT J. HUGGETT, College of William and Mary (retired), Seaford, Virginia
DEBBIE A. NIEMEIER, University of California at Davis
DANIEL NOCERA, Massachusetts Institute of Technology
MICHAEL OPPENHEIMER, Princeton University
DAN REICHER, Stanford University
BERNARD ROBERTSON, DaimlerChrysler Corporation (retired), Bloomfield Hills, Michigan
GARY ROGERS, FEV, Inc., Auburn Hills, Michigan
ALISON SILVERSTEIN, Consultant, Pflugerville, Texas
MARK THIEMENS, University of California at San Diego
RICHARD WHITE, Oppenheimer & Company, New York

Staff

JAMES J. ZUCCHETTO, Director
DANA CAINES, Financial Associate
DAVID COOKE, Associate Program Officer
ALAN CRANE, Senior Scientist
JOHN HOLMES, Senior Program Officer/Associate Director
LaNITA JONES, Administrative Coordinator
ALICE WILLIAMS, Senior Project Assistant
JONATHAN YANGER, Senior Project Assistant

Preface

Recent scientific and technological progress in inertial confinement fusion (ICF), together with the National Ignition Campaign (NIC) to achieve the important milestone of ignition on the National Ignition Facility (NIF), motivated the Department of Energy's (DOE's) Office of the Under Secretary for Science to request that the National Research Council (NRC) undertake a study to assess the prospects for inertial fusion energy (IFE) and provide advice on the preparation of a research and development (R&D) roadmap leading to an IFE demonstration plant. The statement of task for the full NRC study reads as follows:

> The Committee will prepare a report that will:
>
> - Assess the prospects for generating power using inertial confinement fusion;
> - Identify scientific and engineering challenges, cost targets, and R&D objectives associated with developing an IFE demonstration plant; and
> - Advise the U.S. Department of Energy on its development of an R&D roadmap aimed at creating a conceptual design for an inertial fusion energy demonstration plant.

In response to this request, the NRC established the Committee on the Prospects for Inertial Confinement Fusion Energy Systems. As part of the study, the sponsor also requested that the NRC provide an interim report to assist it in formulating its budget request for future budget cycles (see Appendix B). The interim report had a limited scope and was released in March 2012.[1]

---

[1] National Research Council, 2012. *Interim Report—Status of the Study "An Assessment of the Prospects for Inertial Fusion Energy,"* The National Academies Press, Washington, D.C. Available at http://www.nap.edu/catalog.php?record_id=13371.

The committee's final report represents the consensus of the committee after six meetings (see Appendix C for the meeting agendas). The first four meetings were concerned mainly with information gathering through presentations, while the final two meetings focused on carrying out a detailed analysis of the many important topics needed to complete the committee's assessment.

This report describes and assesses the current status of inertial fusion energy research in the United States, identifies the scientific and engineering challenges associated with developing inertial confinement fusion as an energy source, compares the various technical approaches, and, finally, provides guidance on an R&D roadmap at the conceptual level for a national program aimed at the design and construction of an inertial fusion energy demonstration plant, including approximate estimates, where possible, of the funding required at each stage. At the outset of the study, the committee decided that the fusion-fission hybrid concept was outside the scope of the study. While they are certainly interesting subjects of study, comparisons of inertial fusion energy to magnetic fusion energy or any other potential or available energy technologies (such as wind or nuclear fission) were also outside the committee's purview.

Although the committee carried out its work in an unclassified environment, it was recognized that some of the research relevant to the prospects for inertial fusion energy was conducted under the auspices of the nation's nuclear weapons program and has been classified. Therefore, the NRC established the separate Panel on the Assessment of Inertial Confinement Fusion (ICF) Targets to explore the extent to which past and ongoing classified research affects the prospects for practical inertial fusion energy systems. The panel was also tasked with analyzing the nuclear proliferation risks associated with IFE; although that analysis was not available for inclusion in the interim report, the committee reviewed the panel's principal conclusions and recommendations on proliferation, and these are included in this final report of the committee.

The target physics panel exchanged unclassified information informally with the committee in the course of the study process, and the committee was aware of the panel's conclusions and recommendations as they evolved.

The panel produced both a classified and an unclassified report; the latter was timed so as to be available to inform this committee's final report; the Summary of the panel's unclassified report (prepublication version) is included as Appendix H. The statement of task for the panel is given in Appendix B and the panel's meeting agendas appear in Appendix C. The panel's unclassified report, *Assessment of Inertial Confinement Fusion Targets*, is being released simultaneously with this, the committee's final report.

Over the course of the study, the inertial confinement fusion community provided detailed information on the current status and potential prospects for all aspects of IFE. This information and the associated interactions with the community were essential to the committee's work. We, as co-chairs of the committee, recognize the enormous amount of time and effort involved in this contribution and thank the community for its extensive input and help with its task. Finally, we are particularly grateful to the members of this committee who worked so diligently over nearly 2 years to produce this report.

Finally, we would like to express our deep appreciation to the staff at the NRC, particularly to David Lang and Greg Eyring, for their highly professional contributions at every stage of the committee's deliberations and preparation of the report. We are truly indebted to them for their insights and extraordinary contributions throughout the entire process.

> Ronald C. Davidson, *Co-Chair*
> Gerald L. Kulcinski, *Co-Chair*
> Committee on the Prospects for Inertial
> Confinement Fusion Energy Systems

Acknowledgment of Reviewers

This report has been reviewed in draft form by individuals chosen for their diverse perspectives and technical expertise, in accordance with procedures approved by the Report Review Committee of the National Research Council (NRC). The purpose of this independent review is to provide candid and critical comments that will assist the institution in making its published report as sound as possible and to ensure that the report meets institutional standards for objectivity, evidence, and responsiveness to the study charge. The review comments and draft manuscript remain confidential to protect the integrity of the deliberative process. We wish to thank the following individuals for their review of this report:

Douglas M. Chapin, MPR Associates;
Philip Clark, GPU Nuclear Corporation (retired);
Michael I. Corradini, University of Wisconsin;
Todd Ditmire, University of Texas at Austin;
R. Paul Drake, University of Michigan;
Douglas Eardley, University of California at Santa Barbara;
Arjun Makhijani, Institute for Energy and Environmental Research;
Gregory Moses, University of Wisconsin;
Burton Richter, Stanford University;
Robert H. Socolow, Princeton University;
Frank N. von Hippel, Princeton University; and
Steven Zinkle, Oak Ridge National Laboratory.

Although the reviewers listed above have provided many constructive comments and suggestions, they were not asked to endorse the conclusions or recommendations, nor did they see the final draft of the report before its release. The review of this report was overseen by Louis J. Lanzerotti, New Jersey Institute of Technology. Appointed by the NRC, he was responsible for making certain that an independent examination of this report was carried out in accordance with institutional procedures and that all review comments were carefully considered. Responsibility for the final content of this report rests entirely with the authoring committee and the institution.

Contents

SUMMARY 1

1 INTRODUCTION 12
   Overall Power Plant Efficiency, 17
   Drivers, 18
   Targets, 19
   Chambers, 21
   Major Conclusions of Previous Studies, 22
   Major U.S. Research Programs, 23
   Major Foreign Programs, 23
   Statement of Task, 26
   Scope and Committee Approach, 27
   Structure of the Report, 28

2 STATUS AND CHALLENGES FOR INERTIAL FUSION
   ENERGY DRIVERS AND TARGETS 29
   Methods for Driving the Implosion of Targets, 30
   Driver Options for Inertial Confinement Fusion, 45
   General Conclusions, 88

3 INERTIAL FUSION ENERGY TECHNOLOGIES 89
   High-Level Conclusions and Recommendations, 90
   Target Fabrication and Handling for Inertial Fusion Energy, 91

Chamber Technology, 106
Materials, 118
Tritium Production, Recovery, and Management, 125
Environment, Health and Safety Considerations, 129
Balance-of-Plant Considerations, 137
Economic Considerations, 139

4   A ROADMAP FOR INERTIAL FUSION ENERGY   146
    Introduction, 148
    Technology Applications, 150
    Event-Based Roadmaps, 152
    Composite Roadmap and Decision Analysis for the
        Pre-ignition Stage, 156
    TRLs for Inertial Fusion Energy, 161
    Cost and Funding Considerations, 163
    The Need for a National Inertial Fusion Energy R&D Program, 167

APPENDIXES

A	The Basic Science of Inertial Fusion Energy	173
B	Statements of Task	177
C	Agendas for Committee Meetings and Site Visits	179
D	Agendas for Meetings of the Panel on the Assessment of Inertial Confinement Fusion (ICF) Targets	193
E	Bibliography of Previous Inertial Confinement Fusion Studies Consulted by the Committee	197
F	Foreign Inertial Fusion Energy Programs	199
G	Glossary and Acronyms	204
H	Summary from the Report of the Panel on the Assessment of Inertial Confinement Fusion (ICF) Targets (Unclassified Version)	210
I	Technical Discussion of the Recent Results from the National Ignition Facility	219
J	Detailed Discussion of Technology Applications Event Profiles	222

Summary

The potential for using fusion energy to produce commercial electric power was first explored in the 1960s. Harnessing fusion energy offers the prospect of a nearly-carbon-free energy source with a virtually unlimited supply of fuel (it is derived from deuterium in water). Moreover, unlike nuclear fission plants, fusion power plants, if appropriately designed, would not produce large amounts of high-level nuclear waste requiring long-term disposal. These prospects induced many nations around the world to initiate research and development (R&D) programs aimed at developing fusion as an energy source. Two alternative approaches are being explored: magnetic fusion energy (MFE) and inertial fusion energy (IFE). This report assesses the prospects for IFE, although there are some elements common to the two approaches. Recognizing that the practical realization of fusion energy remains decades away, the committee nonetheless judges that the potential benefits of IFE justify it as part of the long-term U.S. energy R&D portfolio.

To initiate fusion, the deuterium and tritium fuel must be heated to over 50 million degrees and held together long enough for the reactions to take place (see Appendix A). Making inertial fusion a commercial source of energy depends on the ability to implode a fuel target to a high enough temperature and pressure to initiate a fusion reaction that releases on the order of 100 times more energy than was delivered to the target.

The current U.S. fleet of inertial fusion facilities offers a unique opportunity to experiment at "fusion scale," where fusion conditions are accessible for the first time. Indeed, significant fusion burn is expected on the National Ignition Facility

(NIF) by the end of this decade. A key aim of this study is to determine how best to exploit this opportunity to advance the science and technology of IFE.

CURRENT R&D STATUS

U.S. research on inertial confinement fusion (ICF)—one of the two ways (the other is magnetic confinement fusion) energy is produced by means of fusion—has been supported by the National Nuclear Security Administration (NNSA), primarily for applications related to stewardship of the nuclear-weapons stockpile. This research has benefited inertial fusion for energy applications because the two share many common physics challenges.

The principal research efforts in the United States are aligned along the three major energy sources for driving the implosion of inertial confinement fusion fuel pellets: (1) lasers, including solid state lasers at the Lawrence Livermore National Laboratory's (LLNL's) NIF and the University of Rochester's Laboratory for Laser Energetics (LLE), as well as the krypton fluoride gas lasers at the Naval Research Laboratory; (2) particle beams, being explored by a consortium of laboratories led by the Lawrence Berkeley National Laboratory (LBNL); and (3) pulsed magnetic fields, being explored on the Z machine at Sandia National Laboratories.

There has been substantial scientific and technological progress in inertial confinement fusion during the past decade. Despite these advances, the minimum technical accomplishment that would give confidence that commercial IFE may be feasible—the ignition[1] of a fuel pellet in the laboratory—has not been achieved as of this writing.[2]

For the first time, a research facility, the NIF[3] at LLNL, conducted a systematic campaign at an energy scale that was projected to be sufficient to achieve ignition. In anticipation of this, the U.S. Department of Energy (DOE) asked the National Research Council (NRC) to review the prospects for inertial fusion energy with the following statement of task:

- Assess the prospects for generating power using inertial confinement fusion;
- Identify scientific and engineering challenges, cost targets, and R&D objectives associated with developing an IFE demonstration plant; and
- Advise the U.S. Department of Energy on its development of an R&D roadmap aimed at creating a conceptual design for an IFE demonstration plant.

---

[1] In this report, ignition is defined as "scientific breakeven," in which the target releases an amount of energy equal to the energy incident upon it to drive the implosion.

[2] As of December 27, 2012.

[3] The NIF, which was designed for stockpile stewardship applications, currently uses a solid-state laser driver and an indirect-drive target configuration.

A comparison of inertial fusion energy to magnetic fusion energy or any other potential or available energy technologies, such as wind or nuclear fission, while it would be a very interesting subject of study, was outside the committee's purview.

The National Ignition Campaign being carried out at the NIF has made significant technical progress during the past year. Nevertheless, ignition has taken longer than scheduled. The results of the experiments performed to date have differed from model projections and are not yet fully understood. It will likely take much more than a year from now to gain a full understanding of the discrepancies between theory and experiment and to make needed modifications to optimize target performance.[4] Box 1.1 in Chapter 1, "Recent Results from the National Ignition Facility," provides a detailed discussion of the most recent NIF results, and Appendix I provides a more technical discussion of this subject.

While the committee considers the achievement of ignition as an essential prerequisite for initiating a national, coordinated, broad-based inertial fusion energy program, it does not believe that the fact that NIF did not achieve ignition by the end of the National Ignition Campaign (September 30, 2012) lessens the long-term technical prospects for IFE. It is important to note that none of the expert committees[5] that reviewed NIF's target performance concluded that ignition would not be achievable at the facility. Furthermore, as the Panel on the Assessment of ICF Target concluded, "So far as target physics is concerned, it is a modest step from NIF scale to IFE scale."[6] A better understanding of the physics of indirect-drive implosions is needed, as well as improved capabilities for simulating them. In addition, alternative implosion modes (laser direct drive, shock ignition, heavy-ion drive, and pulsed-power drive) have yet to be adequately explored. It will therefore be critical that the unique capabilities of the NIF be used to determine the viability of ignition at the million joule energy scale.

As the scientific basis for IFE is better understood—e.g., ignition is achieved or the conditions for ignition are better understood—the path forward for IFE research will diverge from that for NNSA's weapons research program because technologies specific to IFE (e.g., high-repetition-rate driver modules, chamber materials, mass-producible targets) will need to receive a higher priority.

---

[4] NNSA, 2012. *NNSA's Path Forward to Achieving Ignition in the Inertial Confinement Fusion Program: Report to Congress*, December.

[5] DOE Memo by D.H. Crandall to D.L. Cook, "External Review of the National Ignition Campaign," July 19, 2012; National Ignition Campaign Technical Review Committee, "The National Ignition Campaign Technical Review Committee Report, for the Meeting Held on May 30 through June 1, 2012"; NRC, 2013, *Assessment of Inertial Confinement Fusion Targets*, Washington, D.C.: The National Academies Press.

[6] See Overarching Conclusion 1 from *Assessment of Inertial Confinement Fusion Targets*, 2013.

PRINCIPAL CONCLUSIONS AND RECOMMENDATIONS

With substantial input from the community, the committee conducted an intensive review of approaches to IFE—diode-pumped lasers, krypton fluoride lasers, heavy-ion accelerators, pulsed power, as well as indirect drive[7] and direct drive.[8] The committee's principal conclusions and recommendations regarding its assessment of the prospects for IFE are given below. They are grouped thematically under several general topic headings. Additional conclusions and recommendations are contained in the individual chapters. Where there is overlap, the conclusions and recommendations are numbered as they appear in the chapters, to point the reader to the location of more detailed discussion. The recommendations are made in view of the current technical uncertainties and the anticipated long time frame to achieve commercialization of IFE.

Potential Benefits, Recent Progress, and Current Status of Inertial Fusion Energy

Conclusion: The scientific and technological progress in inertial confinement fusion has been substantial during the past decade, particularly in areas pertaining to the achievement and understanding of high-energy-density conditions in the compressed fuel, and in exploring several of the critical technologies required for inertial fusion energy applications—high-repetition-rate lasers and heavy-ion-beam systems, pulsed-power systems, and cryogenic target fabrication techniques. (Conclusion 1 from the Interim Report;[9] Chapters 2 and 3 of this report)

Conclusion: It would be premature to choose a particular driver approach as the preferred option for an inertial fusion energy demonstration plant at the present time. (Conclusion 2 from the Interim Report)

Conclusion: The potential benefits of energy from inertial confinement fusion (abundant fuel, minimal greenhouse gas emissions, and limited high-level radioactive waste requiring long-term disposal) also provide a compelling rationale for including inertial fusion energy R&D as part of the long-term

---

[7] In an indirect-drive target, the driver energy strikes the inner surface of a hollow chamber (the "hohlraum") that surrounds the fuel capsule, exciting X-rays that transfer energy to the capsule.

[8] In a direct-drive target, the driver energy strikes directly on the fuel capsule. The illumination geometry of the driver beams may be oblique—i.e., from diametrically opposite sides, called "polar direct drive"—or spherically symmetric.

[9] NRC, 2012. *Interim Report—Status of the Study "Assessment of the Prospects for Inertial Fusion Energy,"* Washington, D.C.: The National Academies Press.

R&D portfolio for U.S. energy. A portfolio strategy hedges against uncertainties in the future availability of alternatives such as those that arise from unforeseen circumstances. (Conclusion 1-1)

Factors Influencing the Commercialization of Inertial Fusion Energy

Conclusion: The cost of targets has a major impact on the economics of inertial fusion energy power plants. Very large extrapolations are required from the current state of the art for fabricating targets for inertial confinement fusion research to the ability to mass-produce inexpensive targets for inertial fusion energy systems. (Conclusion 3-24)

Conclusion: As presently understood, an inertial fusion energy power plant would have a high capital cost and would therefore have to operate with a high availability. Achieving high availabilities is a major challenge for fusion energy systems. It would involve substantial testing of IFE plant components and the development of sophisticated remote maintenance approaches. (Conclusion 3-23)

Recommendation: Economic analyses of inertial fusion energy power systems should be an integral part of national program planning efforts, particularly as more cost data become available. (Recommendation 3-10)

Recommendation: A comprehensive systems engineering approach should be used to assess the performance of IFE systems. Such analysis should use a Technology Readiness Level (TRL) methodology to help guide the allocation of R&D funds. (Recommendation 3-11)

Conclusion: Some licensing/regulatory-related research has been carried out for the ITER (magnetic fusion energy) program, and much of that work provides insights into the licensing process and issues for inertial fusion energy. The Laser Inertial Fusion Energy (LIFE) program at Lawrence Livermore National Laboratory has considered licensing issues more than any other IFE approach; however, much more effort would be required when a Nuclear Regulatory Commission license is pursued for inertial fusion energy. (Conclusion 3-20)

The Establishment of an Integrated National Inertial Fusion Energy Program and Its Characteristics

Conclusion: While there have been diverse past and ongoing research efforts sponsored by various agencies and funding mechanisms that are relevant to IFE, at the present time there is no nationally coordinated R&D program in the United States aimed at the development of inertial fusion energy that incorporates the spectrum of driver approaches (diode-pumped lasers, heavy ions, krypton fluoride lasers, pulsed power, or other concepts), the spectrum of target designs, or any of the unique technologies needed to extract energy from any of the variety of driver and target options. (Conclusion 4-9)

Conclusion: Funding for inertial confinement fusion is largely motivated by the U.S. nuclear weapons program, because of its relevance to stewardship of the nuclear stockpile. The National Nuclear Security Administration (NNSA) does not have an energy mission and—in the event that ignition is achieved—the NNSA and inertial fusion energy research efforts will continue to diverge as technologies relevant to IFE (e.g., high-repetition-rate driver modules, chamber materials, and mass-producible targets) begin to receive a higher priority in the IFE program. (Conclusion 4-10)

Conclusion: The appropriate time for the establishment of a national, coordinated, broad-based inertial fusion energy program within DOE would be when ignition is achieved. (Conclusion 4-13)

Conclusion: At the present time, there is no single administrative home within the Department of Energy that has been invested with the responsibility for administering a national inertial fusion energy R&D program. (Conclusion 4-16)

Recommendation: In the event that ignition is achieved on the National Ignition Facility or another facility, and assuming that there is a federal commitment to establish a national inertial fusion energy R&D program, the Department of Energy should develop plans to administer such a national program (including both science and technology research) through a single program office. (Recommendation 4-11)

Recommendation: The Department of Energy should use a milestone-based roadmap approach based on technology readiness levels (TRLs) to assist in planning the recommended national IFE program leading to a demonstration plant. The plans should be updated regularly to reassess each potential approach and set priorities based on the level of progress. Suitable milestones for each driver-target pair considered might include, at a minimum, the following technical goals:

1. Ignition,
2. Reproducible modest gain,
3. Reactor-scale gain,
4. Reactor-scale gain with a cost-effective target, and
5. Reactor-scale gain with the required repetition rate. (Recommendation 4-4)

Recommendation: The national inertial fusion energy technology effort should leverage materials and technology development from magnetic fusion energy efforts in the United States and abroad. Examples include ITER's test blanket module R&D program, materials development, plasma-facing components, tritium fuel cycle, remote handling, and fusion safety analysis tools. (Recommendation 3-2)

Inertial Fusion Energy Drivers

Conclusion: Each target design and each driver approach has potential advantages and uncertainties to the extent that "the best driver approach" remains an open question. (Conclusion 4-5)

Laser Drivers

Conclusion: If the diode-pumped, solid-state laser technical approach is selected for the roadmap development path, the demonstration of a diode-pumped, solid-state laser beam-line module and line-replaceable unit at full scale would be a critical step in the development of a laser driver for IFE. (Conclusion 2-2)

Conclusion: If the KrF laser technical approach is selected for the roadmap development path, a very important element of the KrF laser inertial fusion energy research and development program would be the demonstration of a multikilojoule 5- to 10-Hz KrF laser module that meets all of the requirements for a Fusion Test Facility. (Conclusion 2-6)

Heavy-Ion-Beam Drivers

Conclusion: Demonstrating that the Neutralized Drift Compression Experiment-II (NDCX-II) meets its energy, current, pulse length, and spot-size objectives is of great technical importance, both for heavy-ion inertial fusion energy applications and for high-energy-density physics. (Conclusion 2-7)

Conclusion: Restarting the High-Current Experiment to undertake driver-scale beam transport experiments and restarting the enabling technology programs are crucial to reestablishing a heavy-ion fusion program. (Conclusion 2-8)

Pulsed-Power Drivers

Conclusion: There has been considerable progress in the development of efficient pulsed-power drivers of the type needed for inertial confinement fusion applications, and the funding is in place to continue along that path. (Conclusion 2-12)

Conclusion: The major technology development issues that would have to be resolved to make a pulsed-power IFE system feasible—the recyclable transmission line and the ultra-high-yield chamber—are not receiving any significant attention. (Conclusion 2-14)

Recommendation: Physics issues associated with the Magnetized Liner Inertial Fusion (MagLIF) concept should be addressed in single-pulse mode during the next 5 years so as to determine its scientific feasibility. (Recommendation 2-2)

Recommendation: Technical issues associated with the viability of recyclable transmission lines and 0.1-Hz, 10-GJ-yield chambers should be addressed with engineering feasibility studies in the next 5 years to assess the technical feasibility of MagLIF as an inertial fusion energy system option. (Recommendation 2-3)

Other Critical Technologies for Inertial Fusion Energy

Conclusion: Significant IFE technology research and engineering efforts are required to identify and develop solutions for critical technology issues and systems, among them targets and target systems; reaction chambers (first wall/blanket/shield); materials development; tritium production, recovery, and management systems; environment and safety protection systems; and economic analysis. (Conclusion 3-3)

Target Technologies

Conclusion: An inertial fusion energy program would require expanded effort on target fabrication, injection, tracking, survivability, and recycling. Target technologies developed in the laboratory would need to be demonstrated on industrial mass production equipment. A target technology program would be required for all promising inertial fusion energy options, consistent with budgetary constraints. (Conclusion 3-9)

Chamber Technologies

Conclusion: The chamber and blanket are critical elements of an inertial fusion energy power plant, providing the means to convert the energy released in fusion reactions into useful applications as well as the means to breed the tritium fuel. The choice and design of chamber technologies are strongly coupled to the choice and design of driver and target technologies. A coordinated development program is needed. (Conclusion 3-10)

The National Ignition Facility

Conclusion: The National Ignition Facility, designed for stockpile stewardship applications, is also of great potential importance for advancing the technical basis for inertial fusion energy research. (Conclusion 4-15)

Conclusion: There has been good technical progress during the past year in the ignition campaign carried out on the National Ignition Facility. Nevertheless, ignition has been more difficult than anticipated and was not achieved in the National Ignition Campaign, which ended on September 30, 2012. The results of experiments to date are not fully understood. It will likely take significantly more than a year to gain a full understanding of the discrepancies between theory and experiment and to make modifications needed to optimize target performance. (Conclusion 2-1)

Recommendation: The target physics programs on the NIF, Nike, OMEGA, and Z should receive continued high priority. The program on NIF should be expanded to include direct drive and alternative modes of ignition. It should aim for ignition with moderate gain and comprehensive scientific understanding, leading to codes with predictive capabilities for a broad range of IFE targets. (Recommendation 2-1)

Recommendation: The achievement of ignition with laser-indirect drive at the National Ignition Facility should not preclude experiments to test the feasibility of laser-direct drive. Direct-drive experiments should also be carried out because of their potential for achieving higher gain and/or other technological advantages. (Recommendation 4-7)

Recommendation: Planning should begin for making effective use of the National Ignition Facility as one of the major program elements in an assessment of the feasibility of inertial fusion energy. (Recommendation from the Interim Report and Recommendation 4-10 from this report)

Proliferation Risks

The NRC Panel on the Assessment of Inertial Confinement Fusion Targets has examined the proliferation risks associated with IFE systems. Its analysis and principal conclusions regarding proliferation risks are presented in Chapter 3 of its report, *Assessment of Inertial Confinement Fusion Targets*. The NRC Committee on the Prospects for Inertial Confinement Fusion Energy Systems concurs with the Panel's conclusions, which are reiterated below for completeness.

Conclusion: At present, there are more proliferation concerns associated with indirect-drive targets than with direct-drive targets. However, worldwide technology developments may eventually render these concerns moot.[10] Remaining concerns are likely to focus on the use of classified codes for target design. (Conclusion 3-1 from the panel report)

Conclusion: The nuclear weapons proliferation risks associated with fusion power plants are real, but are likely to be controllable.[11] These risks fall into three categories: knowledge transfer; Special Nuclear Material (SNM) production; and tritium diversion. (Conclusion 3-2 from the panel report)

Conclusion: Research facilities are likely to be a greater proliferation concern than power plants. A working power plant is less flexible than a research facility, and it is likely to be more difficult to explore a range of physics problems with a power plant. However, domestic research facilities (which may have a

---

[10] Progress in experiment and computation may eventually result in data, simulations, and knowledge that the U.S. presently considers classified becoming widely available. Classification concerns about different kinds of targets may then change considerably.

[11] Proliferation of knowledge and production of Special Nuclear Material are subject to control by international inspection of research facilities and plants; tritium diversion is a problem that will require careful attention.

mix of defense and scientific missions) are more complicated to put under international safeguards than commercial power plants. Furthermore, the issue of proliferation from research facilities will have to be dealt with long before proliferation from potential power plants becomes a concern. (Conclusion 3-3 from the panel report)

Conclusion: It will be important to consider international engagement regarding the potential for proliferation associated with IFE power plants. (Conclusion 3-4 from the panel report)

1

Introduction

The desirability of fusion power is undeniable. There is, after all, sufficient fusion fuel to supply the entire world's energy needs for millions of years.[1] Furthermore, fusion power plants would have negligible environmental impact since they would produce no greenhouse gases and, if appropriately designed, no long-lived radioactive waste.[2] However, achieving fusion at the cost and scale needed for energy generation is still a major challenge.[3] To initiate fusion, the deuterium and tritium fuel must be heated to over 50 million degrees and held together for long enough for the reactions to take place (see Appendix A). The two main approaches to fusion achieve these conditions differently: In magnetic confinement fusion, the low-density fuel is held indefinitely in a magnetic field while it reacts; in inertial confinement fusion (ICF), a small capsule of fuel (the "target") is compressed and heated so that it reacts rapidly before it disassembles (see Figure 1.1). In this study, the committee assesses the prospects and challenges for generating power using ICF.

The current U.S. fleet of inertial fusion facilities offers a unique opportunity to experiment at "fusion scale," where fusion conditions are accessible for the first

---

[1] Tritium (superheavy hydrogen) and deuterium (heavy hydrogen) are the fuels for the easiest fusion reaction. Tritium must be made by being "bred" from lithium. One liter of sea water contains enough lithium and deuterium to make roughly 1 kWh of fusion energy. See Appendix A.

[2] S.W. White and G.L. Kulcinski, 2000, Birth to death analysis of the energy payback ratio and CO_2 gas emission rates from coal, fission, wind, and DT-fusion electrical power plants, *Fusion Engineering and Design* 48: 473-481.

[3] To initiate fusion, the deuterium and tritium fuel must be heated to over 50 million degrees and held together long enough for the reactions to take place (see Appendix A).

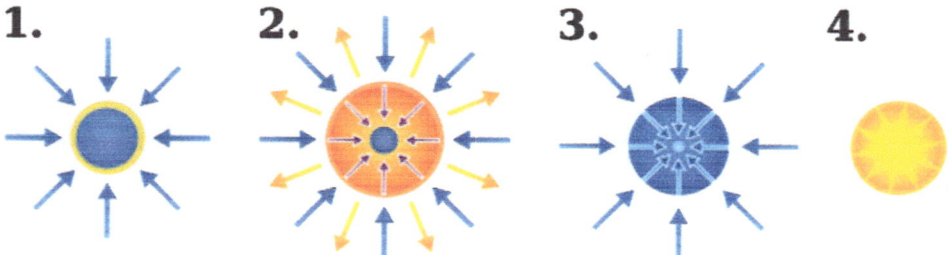

FIGURE 1.1 Simple schematic of the four stages of inertial confinement fusion via hot spot ignition. Stage 1: Energy is delivered to the surface of a tiny hollow sphere a few millimeters in diameter containing fusion fuel (the "target"). The blue arrows represent the "driver energy" delivered to the target—this is the laser light, X-rays, or particle beams that heat the outer yellow shell. Stage 2: Orange arrows indicate the ablation of the outer shell that pushes the inner shell toward the center. The compression of the fusion fuel to very high density increases the potential fusion reaction rate. Stage 3: The central low-density region, comprising a small percentage of the fuel, is heated to fusion temperatures. The light blue arrows represent the energy transported to the center to heat the hot spot. This initiates the fusion burn. Stage 4: An outwardly propagating fusion burn wave triggers the fusion of a significant fraction of the remaining fuel during the brief period before the pellet explodes/disassembles. Steady power production is achieved through rapid, repetitive fusion microexplosions of this kind. (A more detailed primer on the physics is given in Appendix A.)

time. Indeed, significant fusion burn is expected on the National Ignition Facility (NIF) in this decade (see Box 1.1). A key aim of this study is to determine how best to exploit the opportunity offered by the NIF to advance the science and technology of inertial fusion energy (IFE). The committee judges that the potential benefits of IFE justify its inclusion as part of the long-term U.S. energy R&D portfolio, recognizing that the practical realization of fusion energy remains decades away.

> **Conclusion 1-1: The potential benefits of energy from inertial confinement fusion (abundant fuel, minimal greenhouse gas emissions, and limited high-level radioactive waste requiring long-term disposal) also provide a compelling rationale for including inertial fusion energy R&D as part of the long-term R&D portfolio for U.S. energy. A portfolio strategy hedges against uncertainties in the future availability of alternatives, such as those that arise from unforeseen circumstances.**

While the IFE concept is simple, the practical implementation and the high-energy-density target physics are not. If the compression of the target is insufficient, the fusion reaction rate is too slow and the target disassembles before the reactions take place. Delivering the driver energy and compressing the target

BOX 1.1
Recent Results from the National Ignition Facility

The National Ignition Campaign (NIC) formally ended on September 30, 2012, but the effort to achieve thermonuclear ignition on the NIF is expected to continue, albeit at a somewhat reduced level. While the initial expectations of LLNL scientists for a speedy success in achieving ignition were dashed, much progress was made toward the goal of demonstrating thermonuclear ignition in the laboratory for the first time. The NIC experimental plan for cryogenic deuterium-tritium (DT) layered target implosions and diagnostics is described in the reference given in the footnote.[1] The latest results on the implosion performance are provided in S.H. Glenzer et al.[2] Future directions for experimental and theoretical investigations are described in the proceedings of the workshop on the science of ignition.[3]

Experts in high-energy-density science and ICF convened in San Ramon, California, between May 22 and 24, 2012, for the international workshop Science of Fusion Ignition on NIF to review the results of the NIC experiments, so as to identify major science issues and propose priorities for future research to enhance the understanding of ignition in ICF. Subpanels of specialists analyzed results in all of the areas relevant to the implosion physics, from laser-plasma interaction and radiation transport, to implosion hydrodynamics, and burn physics. In their final report, the group of experts recognizes the need for an improved predictive capability to better guide ignition experiments. They recommend specific experiments to validate models and codes, and to improve basic understanding of the complex physics phenomena occurring in a laser-driven implosion.

In their most recent review, on May 31, 2012, a team appointed by the NNSA also concluded that "better understanding through detailed measurements and model adjustments informed by rigorous quantifications of uncertainties are needed both to better approach the ignition process and to benefit the stockpile stewardship program."[4] Another review panel, the NIC Technical Review Committee, concluded that "the NIF is operating in a stable, reliable, predictable, and controllable manner" and that "there is sufficient body of knowledge regarding nuclear fusion and plasma physics to conclude that it should be possible to achieve controlled thermonuclear fusion on a laboratory scale."[5] NNSA recently released a report that lays out a 3-year plan for NNSA's ICF program, stating that "the emphasis going forward will be to illuminate the physics and to improve models and codes used in the ICF program until agreement with experimental data is achieved. Once the codes and models are improved to the point at which agreement is reached, NNSA will be able to determine whether and by what approach ignition can be achieved at the NIF."[6]

An overall performance parameter used by the LLNL group is the experimental ignition threshold factor (ITFx).[7] The ITFx has been derived by fitting the results of hundreds of computer simulations of ignition targets to find a measurable parameter indicative of the performance with respect to ignition. An implosion with ITFx = 1 has a 50 percent probability of ignition. To date, the highest value of the ITFx achieved in DT layered implosion experiments on NIF is about 0.1.[8] To improve the implosion performance and raise the ITFx the LLNL group is taking several steps to reduce the ablator-fuel mix. Further reducing target surface roughness[9] is an obvious remedy. Other available options range from a thicker ablator, a thicker ice layer, and higher entropy implosions. All of these options come with a laser energy penalty. To drive thicker ice or thicker ablator targets will require more laser energy to reach the required implosion velocity. Higher entropy implosions will be more hydrodynamically stable, but high entropy degrades the areal density thus reducing both the one-dimensional margin for ignition and the energy gain in the event of ignition. Another possible cause of performance degradation is the growth of long wavelength spatial nonuniformities induced by asymmetries in the x-ray drive (or other sources).[10] Attempts to mitigate ablator-fuel mix and to measure drive asymmetries are currently under way at LLNL.[11] Other strategies to improve the performance include using different ablators other than plastic (CH). For instance, studies involving high-density carbon or beryllium ablators are under way.

Improving the ignition threshold factor by an order of magnitude will be challenging, but several options are available to improve implosion performance. The continuing experimental campaign at the NIF will explore these options and develop a more fundamental understanding of the key physics issues that are currently preventing the achievement of ignition.

While the committee considers the achievement of ignition as an essential prerequisite for initiating a national, coordinated, broad-based IFE program, the committee does not believe that the fact that NIF did not achieve ignition by the end of the NIC on September 30, 2012, lessens the long-term technical prospects for inertial fusion energy. It is important to note that none of the expert committees[12] that reviewed NIF's target performance concluded that ignition would not be achievable at the facility. Furthermore, as the ICF Target Physics Panel concluded, "So far as target physics is concerned, it is a modest step from NIF scale to IFE scale."[13] A better understanding of the physics of indirect-drive implosions is needed, as well as improved capabilities for simulating them. In addition, alternative implosion modes (laser direct drive, shock ignition, heavy-ion drive, and pulsed power drive) have yet to be adequately explored. It will therefore be critical that the unique capabilities of the NIF be used to determine the viability of ignition at the million joule energy scale.

NOTE: Appendix I provides a technical discussion of the recent results from the NIF.

[1] M.J. Edwards et al., 2011, The experimental plan for cryogenic layered target implosions on the National Ignition Facility—The inertial confinement approach to fusion, *Physics of Plasmas* 18: 051003.

[2] S.H. Glenzer et al., 2012, Cryogenic thermonuclear fuel implosions on the National Ignition Facility, *Physics of Plasmas* 19: 056318.

[3] LLNL, 2012, *Science of Fusion Ignition on NIF*, Report from the Workshop on the Science of Fusion Ignition on NIF held on May 22-24, Document LLNL-TR-570412; available at http://tinyurl.com/8p879e6.

[4] DOE, 2012, Memo by D.H. Crandall to D.L. Cook, "External Review of the National Ignition Campaign," July 19.

[5] NIC Technical Review Committee, "The National Ignition Campaign Technical Review Committee Report, For the Meeting Held on May 30 through June 1, 2012."

[6] NNSA, 2012, *NNSA's Path Forward to Achieving Ignition in the Inertial Confinement Fusion Program: Report to Congress*, December.

[7] B.K. Spears et al., 2012, Performance metrics for inertial confinement fusion implosions: Aspects of the technical framework for measuring progress in the National Ignition Campaign, *Physics of Plasmas* 19: 056316.

[8] S.H. Glenzer et al., 2012, Cryogenic thermonuclear fuel implosions on the National Ignition Facility, *Physics of Plasmas* 19: 056318; and R. Betti, 2012, "Theory of Ignition and Hydroequivalence for Inertial Confinement Fusion, Overview Presentation," OV5-3, 24th IAEA Fusion Energy Conference, October 7-12, San Diego, Calif.

[9] NIC Technical Review Committee, "The National Ignition Campaign Technical Review Committee Report, For the Meeting Held on May 30 through June 1, 2012."

[10] Ibid.

[11] Ibid.

[12] DOE, Memo by D.H. Crandall to D.L. Cook, "External Review of the National Ignition Campaign," July 19, 2012; NIC Technical Review Committee, "The National Ignition Campaign Technical Review Committee Report, For the Meeting Held on May 30 through June 1, 2012"; National Research Council, "Assessment of Inertial Confinement Fusion Targets," The National Academies Press, Washington, D.C., 2012.

[13] See Overarching Conclusion 1 from the Panel report, *Assessment of Inertial Confinement Fusion Targets*, released as a prepublication in early 2013.

uniformly without exciting instabilities that compromise the compression requires high precision in space and timing. Large capsules/targets are in many ways easier since they disassemble more slowly and therefore require less compression. They can also deliver greater gain ("gain" is fusion energy out divided by the driver energy delivered to compress and heat the capsule). However, the fusion energy per explosion—and therefore the size of the capsule—is limited by the need to contain and utilize the energy released. Thus capsules with yields of approximately 100 MJ to 10 GJ (the latter is equivalent to the explosive power of 2.5 tons of TNT) have been proposed as candidates for energy production. The issues that influence the technology choices are explored in subsequent chapters. High fusion gain with limited yield is a prerequisite for practical IFE.

An IFE power plant must do much more than simply ignite a high-gain target. Commercial power production requires many integrated systems, each with technological challenges. It must make the targets, ignite targets repetitively, extract the heat, breed tritium from lithium (see Appendix A), and generate electricity. Furthermore it must do this reliably and economically. The fully integrated system (see Figure 1.2) consists of (1) a target factory to produce about 10^7 to 10^9 low-cost targets per year, (2) a driver to heat and compress the targets to ignition, (3) a fusion chamber to recover the fusion energy pulses from the targets and breed the

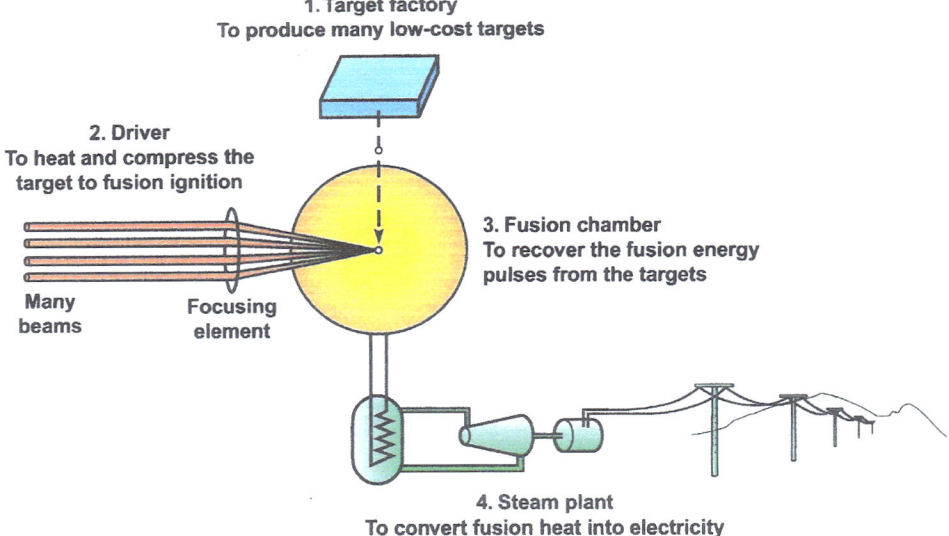

FIGURE 1.2 Schematic of the four major components of an IFE power plant. SOURCE: Opportunities in the Fusion Energy Sciences Program, 1999, http://www.ofes.fusion.doe.gov/more_html/FESAC/FES_all.pdf.

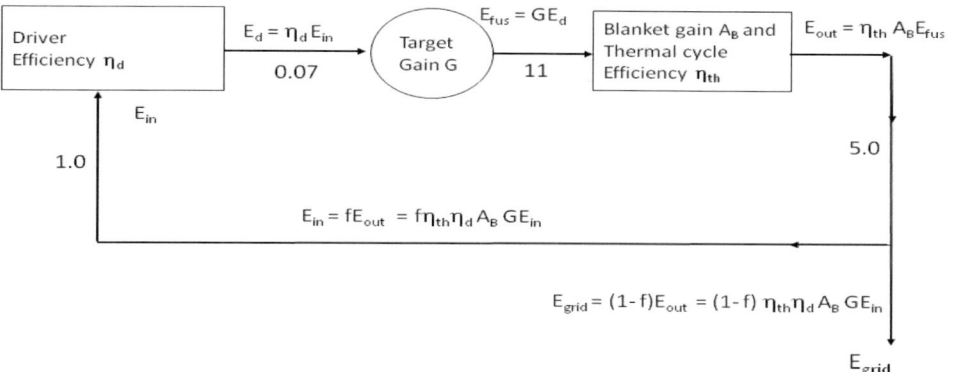

FIGURE 1.3 Schematic energy flow in an inertial fusion power plant. Note that $Q_E = 1/f$. The numbers beside the arrows indicate the proportionality of the energy flows. Tritium breeding (discussed in Chapter 3) is excluded from this diagram for simplicity.

tritium, and (4) a steam plant to convert fusion heat into electricity.[4] A key goal for exploring the engineering feasibility of IFE will be to achieve reproducible gain at the required repetition rate.

OVERALL POWER PLANT EFFICIENCY

Although target gain can be used to validate the target physics, a new parameter is required for assessing the viability of a fusion energy system. The so-called "engineering Q," or "Q_E," is often used as a figure of merit for a power plant. It represents the ratio of the total electrical power produced to the (recirculating) power required to run the plant—that is, the input to the driver and other auxiliary systems. $Q_E = 1/f$, where f is the recycling power fraction (see Figure 1.3). Typically, $Q_E \geq 10$ is required for a viable electrical power plant. For a power plant with a driver wall-plug efficiency η_D, target gain G, thermal-to-electrical conversion efficiency η_{th}, and blanket amplification A_B,[5] $Q_E = \eta_{th}\eta_D A_B G$ (see Figure 1.3). Achievable values of the blanket amplifications and thermal efficiency might be $A_B \sim 1.1$ and $\eta_{th} \sim 0.4$ and should be largely independent of the driver. Therefore,

---

[4] W. Meier, F. Najmabadi, J. Schmidt, and J. Sheffield, "Role of Fusion Energy in a Sustainable Global Energy Strategy," 18th World Energy Congress, Buenos Aires, Argentina, March 7, 2001. Available at http://tinyurl.com/ck84fao.

[5] Amplification, A_B, is the energy multiplier—a dimensionless number—on the total energy of 14.1 MeV neutrons entering the blanket via nuclear reactions with the structural, coolant, and breeding material.

the required target gain is inversely proportional to the driver efficiency. For a power plant with a large recirculating power f = 20 percent (Q_E = 5), the required target gain is G = 75 for a 15 percent efficient driver, and G = 160 for a 7 percent efficient driver.

There will likely be some shot-to-shot variation in target gain resulting from imperfect fabrication, variations in driver pulses, and fluctuations in beam alignment. A power plant must even allow for the possibility of some complete duds. An important goal of the program will be to achieve very good reproducibility and to increase the average target gain as close as possible to the best achievable value (see Table 1.1). In this report, the gain values in various tables and milestones are understood to be average reproducible values. For example, where the report lists modest gain as a milestone, the intended meaning is average, reproducible modest gain. Similarly, the ignition milestone includes the requirement of some reproducibility. Ignition on every shot is not likely, particularly initially, but to achieve the ignition milestone, ignition must be demonstrated in multiple cases.

DRIVERS

The driver is required to deliver megajoules of energy in a few nanoseconds—typically, a significant fraction of a petawatt of power. This energy must be delivered with an electrical efficiency η_D of around 10 percent or more. Four main systems are being studied as potential drivers of inertial fusion plants: diode-pumped, solid-state lasers (DPSSLs), krypton fluoride (KrF) gas lasers, heavy-ion beams from accelerators, and pulsed (electric) power drivers that are connected directly to a load that contains the target. See Chapter 2 for a full description of these options.

TABLE 1.1 Some Reference Examples of Driver, Target, and Chamber Wall Options

Driver	Electrical Efficiency η_D (%)	Energy (MJ)/ Repetition Rate (Hz)	Target Type	Target Gain G	Chamber Wall
DPSS laser	16	1.8-2.2/16	Indirect	60-90	Solid
KrF laser	7	0.5-2.0/10	Direct	100-250	Solid
Heavy ion	25-45	1.8-3.3/5	Indirect	90-130	Liquid
Pulsed power	20-50	33/0.1	Magnetic direct	~300	Liquid

NOTE: Many other examples are possible; their validation will require confirmation from the NIF or other experimental facilities. These figures represent values that are hoped to be achievable. It has not yet been demonstrated that these driver energies are sufficient to achieve ignition and the indicated gain with current implosion parameters. These examples used computations of different levels of sophistication.
SOURCE: Presentations to the committee and their supporting papers.

TARGETS

Current inertial confinement fusion (ICF) targets are made by hand, which is time consuming and expensive. For commercial viability, these high-precision targets must be mass-produced cheaply. Proposed targets vary, depending on the driver, from yields of ~100 MJ to 10 GJ, and the price required for commercial viability depends on many factors. To set the typical scale, consider a plant with a repetition rate of 10 targets per second and 1 GW electrical output; with typical thermal efficiencies, this would mean a target yield of approximately 250 MJ. The cost of targets will depend on many factors, including their materials, complexity, and yield. It is estimated that the fraction of the cost of electricity from an IFE power plant that the manufacturing of targets contributes will range from about 6 percent for the relatively simpler direct-drive laser targets to more than 30 percent for the more complex indirect-drive laser targets, with heavy-ion fusion and pulsed-power targets falling between these two.[6,7,8] IFE target masses are small (usually less than 1 g) and the cost of materials is minimal unless gold or other expensive elements are used. Therefore, the challenge for IFE is the development of manufacturing techniques that can achieve the required cost and precision (see Chapter 3).[9]

For laser-driven fusion, targets come in two main categories: direct-drive targets, in which the driver energy is coupled directly into the target; and indirect-drive targets, in which the driver energy is used to make X-rays inside a cavity called a hohlraum that couple to the target (see Figure 1.4). For heavy-ion and pulsed-power fusion, the distinction between direct and indirect drive is not as clear, as discussed in more detail in Chapter 2. To provide the energy that heats the hot spot to initiate fusion burn, several variants—for example, fast ignition or shock ignition on the scheme depicted in Figure 1.1 have been proposed that may yield higher gain (see further discussion in Chapter 2).

For pulsed-power fusion schemes, tens of millions of amperes of electrical current are pulsed through an assembly around the target. The magnetic pressure created by these currents compresses the target and drives the fusion (see Chapter 2).

---

[6] This percentage includes the fusion fuel (target materials and fabrication costs), the tritium plant, and target injection and tracking. Most of the contribution comes from the target materials and fabrication.

[7] T. Anklam, Lawrence Livermore National Laboratory, "LIFE Economics and Delivery Pathway," Presentation to the committee on January 29, 2011.

[8] D. Goodin, General Atomics, "Target Fabrication and Injection Challenges in Developing an IFE Reactor," Presentation to the committee on January 29, 2011.

[9] W. Meier, F. Najmabadi, J. Schmidt, and J. Sheffield, "Role of Fusion Energy in a Sustainable Global Energy Strategy," 18th World Energy Congress, Buenos Aires, Argentina, March 7, 2001. Available at http://tinyurl.com/ck84fao.

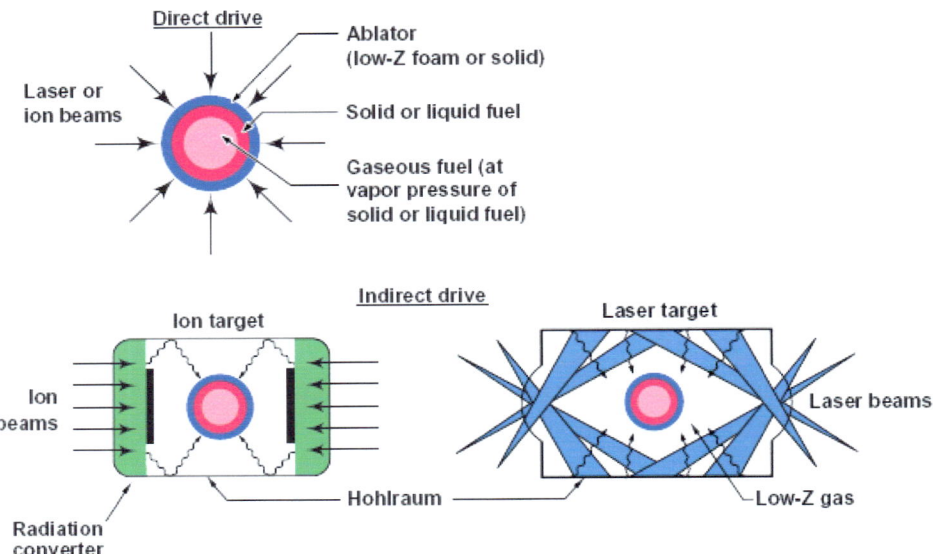

FIGURE 1.4 Direct- and indirect-drive targets. *Top:* Direct-drive target: laser or ion beam shines directly onto the target. *Bottom left:* Ion beam indirect-drive target: ion beams shine on radiation convertor; X-rays (squiggly lines) from radiation convertors fill the inside of the hohlraum and heat the capsule. *Bottom right:* Laser beam indirect-drive target: laser beams shine on the inside of the hohlraum, creating X-rays (squiggly lines) inside the hohlraum that heat the capsule. SOURCE: DOE, Fusion Energy Sciences Committee, "Summary of Opportunities in the Fusion Energy Sciences Program, June 1999." Available at http://tinyurl.com/c4yvffw.

Some of the physics processes involved in ICF for energy applications have parallels with the processes that take place inside thermonuclear weapons, and for this reason most of the research into ICF in the United States has been funded by weapons programs. In modern thermonuclear weapons, a boosted fission device consisting of a plutonium shell containing deuterium and tritium is imploded by conventional explosives. The X-rays produced by the resulting reactions are used to compress a second component. This second component, the "secondary," contains lithium deuteride. The neutrons produced by the reaction D + D are captured in the lithium, producing tritium. The equivalent of up to 60 million tons of high explosives has been released by this process. The IFE effort seeks to release this fusion energy by compression and heating of a small spherical target containing fusion fuel, without the need for a fission trigger.

Because of the parallels between ICF for energy applications and for weapons applications, concerns have been raised about whether pursuit of IFE around the world might facilitate the proliferation of nuclear weapons and expertise. This important issue is discussed in the report of the Panel on the Assessment of Inertial Confinement Fusion (ICF) Targets (see Appendix H for that report's Summary).

CHAMBERS

The fusion reaction yields kinetic energy, one-fifth of which is invested in a helium nucleus (alpha particle) and four-fifths in a neutron (see Appendix A). The alpha particle heats the fuel and supports the burn. Ultimately, however, the alpha energy is emitted as fast-charged particles and X-rays from the exploding capsule. The neutrons barely interact with the capsule and therefore deposit their energy in the chamber wall. Tritium will be bred by the capture of fusion neutrons in lithium—either in a flowing liquid wall of lithium, lithium-lead, or a lithium salt, or in a blanket that contains lithium as a liquid or solid. The energy of the neutrons, the lithium reactions, and the charged particles must all be collected in the chamber walls and used to power a turbine. The tritium must also be collected for use in new capsules.

Making a reliable, long-lived chamber is challenging since the charged particles, target debris, and X-rays will erode the wall surface and the neutrons will embrittle and weaken the solid materials. Many concepts for chamber components have been considered in design studies, including (1) chambers with thick layers of liquid or granules, which protect the structural wall from neutrons, X-rays, charged particles and target debris; (2) first walls that are protected from X-rays and target debris by a thin liquid layer; and (3) dry wall chambers, which are filled with low-pressure gas to protect the first wall from X-rays and target debris. The last two types have structural first walls that must withstand the neutron flux.[10]

Although the specific issues for any particular chamber depend on the choice of driver and target, as well as the choice of wall protection concept, there is a set of challenges that is generic to all concepts: (1) wall protection; (2) chamber dynamics and achievable clearing rate following capsule ignition and burn; (3) injection of targets into the chamber environment; (4) propagation of beams to the target; (5) entry of driver beams into the chamber and protection of the driver from damage; (6) coolant chemistry, corrosion, wetting, and tritium recovery; (7) neutron damage to solid materials; and (8) safety and environmental impacts of first wall, hohlraum, and coolant choices.[11,12]

---

[10] C. Baker, University of California at San Diego, "Advances in Fusion Technology," January 2000, Document UCSD-ENG-077. Available at http://aries.ucsd.edu/LIB/REPORT/UCSD-ENG/UCSD-ENG-077.pdf.

[11] Ibid.

[12] Items (4) and (5) do not apply to pulsed-power IFE.

Many of the issues for inertial fusion regarding materials, the technology of heat exchange, blankets, and tritium recovery are shared with magnetic confinement fusion. Indeed ITER[13] will test breeding blanket modules for the first time. The balance-of-plant (see Chapter 3) will likely be similar to that of existing fission reactors.

MAJOR CONCLUSIONS OF PREVIOUS STUDIES[14]

Over the past 25 years, several prominent studies have reported favorably on scientific progress toward ICF ignition and the prospects for IFE[15] and have recommended that a modest, coordinated program should be initiated that is devoted to energy applications with some level of research on all of the components of an IFE system.[16]

The current designs for IFE plants have used best-guess cost estimates for components and targets.[17] These estimates have provided cost numbers that could be competitive with other future energy sources if there are no major surprises in the physics and technology performance of IFE systems. Chapter 3 provides further discussion of these studies and the economic challenges associated with making IFE a practical energy source.

---

[13] ITER is an international project to build an experimental magnetic confinement fusion reactor in the south of France. It is based on the tokamak concept.

[14] See bibliography in Appendix E.

[15] See, for example, Fusion Policy Advisory Committee (FPAC), 1990, *Final Report*, September; FEAC, Report of the Inertial Fusion Energy Review Panel: July 1996, *Journal of Fusion Energy* 18 (4) 1999; FESAC, 2004, *A Plan for the Development of Fusion Energy*, March.

[16] FEAC, 1994, Panel 7 report on inertial fusion energy, *Journal of Fusion Energy* 13 (2/3); FESAC, 2004, *Review of the Inertial Fusion Energy Research Program*, March.

[17] Examples of such estimates are contained in the following: T.M. Anklam, M. Dunne, W.R. Meier, S. Powers, and A.J. Simon, 2011, LIFE: The case for early commercialization of fusion energy, *Fusion Science and Technology* 60: 66; W.R. Meier, 2008, Systems modeling for a laser-driven IFE power plant using direct conversion, *Journal of Physics Conference Series* 112: 032036; S.S. Yu, W.R. Meier, R.P. Abbott, J.J. Barnard, T. Brown, D.A. Callahan, C. Debonnel, P. Heitzenroeder, J.F. Latkowski, B.G. Logan, S.J. Pemberton, P.F. Peterson, D.V. Rose, G-L. Sabbi, W.M. Sharp, and D.R. Welch, 2003, An updated point design for heavy ion fusion, *Fusion Science and Technology* 44: 266-273; W.R. Meier, 2006, Systems modeling for Z-IFE power plants, *Fusion Engineering and Design* 81: 1661; W.R. Meier, 1994, Osiris and SOMBRERO inertial fusion power plant designs—Summary, Conclusions and Recommendations, *Fusion Engineering and Design* 25: 145-157; L.M. Waganer, 1994, Innovation leads the way to attractive inertial fusion energy reactors—Prometheus-L and Prometheus-H, *Fusion Engineering and Design* 25: 125-143.

MAJOR U.S. RESEARCH PROGRAMS

Inertial fusion energy research gained impetus in the United States following the end of underground nuclear weapons testing in the early 1990s. As a result, major research facilities were constructed to test the physics of target implosion in the laboratory. The work in ICF is funded by the National Nuclear Security Administration (NNSA) and involves the weapons laboratories—Lawrence Livermore National Laboratory (LLNL), Los Alamos National Laboratory (LANL) and Sandia National Laboratories (SNL)—along with the Naval Research Laboratory (NRL) and a number of universities, notably the Laboratory for Laser Energetics (LLE) at the University of Rochester. The major facilities are the lasers NIF at LLNL, OMEGA at LLE, and NIKE at NRL, and the pulsed power system Z at SNL (Box 1.2). The weapons laboratories and a number of universities house smaller facilities. The heavy-ion fusion (HIF) program is undertaken by a Virtual National Laboratory consisting of Lawrence Berkeley National Laboratory (LBNL), LLNL, and the Princeton Plasma Physics Laboratory (PPPL); its present work is focused on high-energy-density physics. The magnetized target fusion approach (see Chapter 2) is studied by LANL and the Air Force.

Sources of funding for IFE R&D have been diverse. They have included Laboratory Directed Research and Development (LDRD) funds at NNSA laboratories—for example, Laser Inertial Fusion Energy (LIFE) and pulsed power approaches—direct funding through the Office of Fusion Energy Sciences—for example, heavy ion fusion, fast ignition, and magnetized target fusion—and congressionally mandated funding. Beginning in FY1999, Congress directed the initiation of the High Average Power Laser (HAPL) program, to be sponsored by NNSA. The HAPL program was an integrated program to develop the science and technology for fusion energy using laser direct drive. Initially focused on the development of solid-state and KrF laser drivers, the program then expanded to address all of the key components of an IFE system, including target fabrication, target injection and engagement, chamber technologies and final optics, and tritium processing. The HAPL program was terminated after FY2009.

MAJOR FOREIGN PROGRAMS

A brief summary of the main foreign IFE programs is given below. A more detailed description can be found in Appendix F.

- *China.* The present program is focused on the development of diode-pumped, solid-state lasers and fast ignition. The near-term goal is fusion ignition and plasma burning, to be achieved around 2020. China is also investigating the use of KrF lasers.

> **BOX 1.2**
> **Major Inertial Confinement Fusion Facilities in the United States**
>
>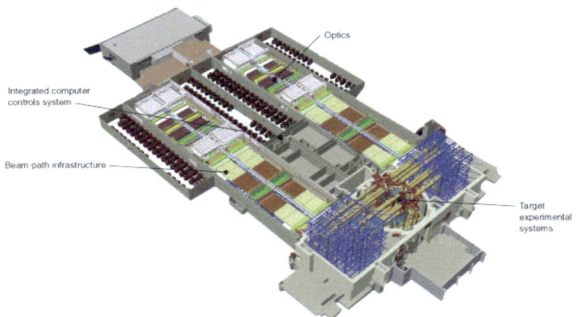
>
> (A) Cutaway illustration of the NIF at LLNL. SOURCE: LLNL, Preparing for the X games of science, *Science & Technology Review*. Available at http://tinyurl.com/7d57jha.
>
>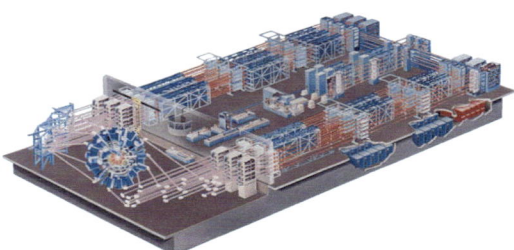
>
> (B) Cutaway illustration of the OMEGA laser facility at the LLE at the University of Rochester. Available at http://tinyurl.com/d57ruq2.
>
>
>
> (C) The Z Pulsed Power Facility at SNL. Available at http://www.sandia.gov/z-machine/.

(D) The NIKE laser target chamber at the NRL. SOURCE: S. Obenschain, NRL, Presentation to the committee on January 29, 2011.

(E) The Neutralized Drift Compression Experiment II (NDCX-II) at LBNL. SOURCE: Roy Kaltschmidt, LBNL. Available at http://tinyurl.com/8xz9kfw.

- *Europe.* The main European Union laser fusion research facilities are in France (the Laser Megajoule (LMJ), the Laboratoire pour l'Utilization des Lasers Intenses (Luli), and Petula); the Czech Republic (Prague Asterix Laser System (PALS)); and the United Kingdom (ORION, Vulcan). The high power laser energy research facility (HiPER) is a power plant study involving 12 countries, including Russia, and is led by the United Kingdom. Its goal is to develop a strategic route to laser fusion power production for Europe. Defining features of HiPER include the high repetition rate; the fact that it is system driven rather than physics driven; and the international, collaborative approach. The present design study envisages using DPSSLs, polar drive, shock ignition (possible test in LMJ at one-third of its maximum energy delivery), and a dry wall with some protection. The start of reactor design is planned for 2026 and operation for 2036. Much of the design of European approaches to IFE is being done using DUED,[18] a code developed in Italy, and MULTI,[19] a code developed in Spain.
- *Germany.* German laboratories are involved in HiPER. Heavy-ion fusion is studied at GSI-Darmstadt using RF-accelerators.
- *Japan.* The main program is focused on DPSSLs and fast ignition with the facility FIREX-1 in operation and FIREX-2 in design. The main goal is for demonstration to begin in 2029. There is collaboration with European programs. A more modest heavy-ion fusion program is undertaken in universities.
- *Russia.* Russia collaborates closely with Germany. The Institute for Theoretical and Experimental Physics' terawatt accumulator (ITEP-TWAC) project will be the main test bed and is now under construction. Russia has recently announced a project to build a 2.8 MJ laser for ICF and weapons research. The Research Institute of Experimental Physics will develop the concept.

STATEMENT OF TASK

Recent scientific and technological progress in ICF, together with the campaign for achieving the important milestone of ignition on the NIF, motivated the DOE's Office of the Under Secretary for Science to request that the National Research Council (NRC) undertake a study that assesses the prospects for IFE, and provides advice on the preparation of an R&D roadmap leading to an IFE demonstration

---

[18] S. Atzeni, A. Schiavi, F. Califano, F. Cattani, F. Cornolti, D. Del Sarto, T.V. Liseykina, A. Macchi, and F. Pegoraro, 2005, Fluid and kinetic simulation of inertial confinement fusion plasmas, *Proceedings of the Europhysics Conference on Computational Physics 2004* 169: 153-159.

[19] R. Ramis, R. Schmalz, and J. Meyer-ter-Vehn, 1988, MULTI—A computer code for one-dimensional multigroup radiation hydrodynamics, *Computer Physics Communications* 49 (3): 475-505.

INTRODUCTION

plant. In response to this request, the NRC established the Committee on the Prospects for Inertial Confinement Fusion Energy Systems; the committee membership is provided in the front matter of this report. The statement of task for the study is as follows:

> The Committee will prepare a report that will:
>
> - Assess the prospects for generating power using inertial confinement fusion;
> - Identify scientific and engineering challenges, cost targets, and R&D objectives associated with developing an IFE demonstration plant; and
> - Advise the U.S. Department of Energy on its development of an R&D roadmap aimed at creating a conceptual design for an inertial fusion energy demonstration plant.
>
> The Committee will also prepare an interim report to inform future year planning by the federal government.

SCOPE AND COMMITTEE APPROACH

The study committee, consisting of 22 members from many fields, published its interim report in 2012.[20] Although the committee carried out its work in an unclassified environment, it was recognized that some of the research relevant to the prospects for IFE systems has been conducted under the auspices of the nation's nuclear weapons program and has been classified. Therefore, the NRC established a separate Panel on the Assessment of Inertial Confinement Fusion (ICF) Targets to explore the extent to which past and ongoing classified research affects the prospects for practical inertial fusion energy systems. The panel was also tasked with the analysis of the nuclear proliferation risks associated with IFE. The panel's statement of task is given in Appendix B.

The panel on targets exchanged unclassified information informally with the committee in the course of the study process, and the committee was aware of its evolving conclusions. The unclassified version of the Summary from the panel's report is included as Appendix H.

The analysis in this report is based on the following:

- Reviewing many past studies on inertial fusion energy systems (see Appendix E);
- Receiving briefings on ongoing research related to IFE systems in the United States and around the world;

---

[20] NRC, 2012, *Interim Report—Status of the Study "Assessment of the Prospects for Inertial Fusion Energy,"* The National Academies Press: Washington, D.C.

- Conducting site visits to major inertial confinement fusion facilities in the United States; and
- Exploiting the expertise of its membership in key areas relating to inertial confinement fusion.

The committee held seven meetings and four site visits at which presentations were invited from key researchers (both national and international) in the field, skeptics who question the current approaches, and independent experts in areas relevant to the commercialization of new technologies. At each meeting, there was also opportunity for public comment. Meeting agendas are given in Appendix C. During the course of the study, the committee consulted with most of the key individuals and laboratories at the forefront of IFE-related research.

STRUCTURE OF THE REPORT

Chapter 2 describes the status of the main approaches to driving the implosion of IFE targets as well as specific challenges that must be met in the near term, medium term, and far term to make the various drivers suitable for use in commercial IFE plants. The status and R&D challenges of the targets themselves, as well as those of the other components of an IFE plant, are discussed in Chapter 3, which also includes a discussion of economic considerations associated with the commercialization of IFE. Finally, Chapter 4 describes the committee's proposed R&D roadmaps for various driver-target combinations in the form of branching decision trees leading to an IFE demonstration plant, as required in its statement of task. For each technological approach, the committee identifies a series of critical R&D objectives that must be met for that approach to be viable. If these objectives cannot be met, then other approaches will need to be considered.

2

Status and Challenges for Inertial Fusion Energy Drivers and Targets

A brief introduction to the concepts of drivers, targets, and implosion mechanisms was given in Chapter 1. In the first part of this chapter, the committee provides a more detailed discussion of alternative strategies for driving the implosion of targets and explains why terms such as "direct drive" and "indirect drive" are more accurate descriptors for some driver-target pairs than for others.

In the second part of this chapter, the committee takes up the status and future R&D needs of the three main driver candidates: lasers (which include diode-pumped, solid-state lasers and KrF lasers); heavy-ion accelerators; and pulsed-power drivers. This discussion of driver approaches is based on input received from proponents who are technical experts in the field.[1] As such, the R&D challenges and investment priorities for moving each approach forward to a major test facility—Fusion Test Facility (FTF)—are discussed independently of one another—that is, as if a decision had been made to choose that particular approach as the best option for inertial fusion energy (IFE). The committee recognizes that a down-selection to one particular approach will have to be made and does not mean to suggest that all of the approaches should be funded simultaneously at the levels indicated in this chapter. A discussion of how these approaches might fit into an integrated program with down-selection decision points is given in Chapter 4. Throughout this chapter material is drawn from the report of the committee's supporting Target Physics Panel (see the Preface); the Summary from the unclassified Target Physics Panel report appears as Appendix H.

---

[1] The experts who gave presentations to the committee are listed in Appendix C.

Conclusions and recommendations are given within the sections. General conclusions appear at the end of this chapter.

METHODS FOR DRIVING THE IMPLOSION OF TARGETS

A large number of target designs have been studied and proposed for IFE power plants. As explained in Chapter 1, these targets may be categorized according to the method used to drive the implosion (to compress the fuel to high density) and according to the method used to bring the fuel to the required ignition temperature. In addition, targets are sometimes categorized according to illumination geometry. For example, in some target designs, the incoming driver beams are arranged uniformly around the target to approximate spherical illumination. At the National Ignition Facility (NIF), the beams are arranged in four cones that illuminate the inside wall of the hohlraum from two sides (the poles of the cylindrically symmetric target). Historically, there have also been illumination geometries that more strongly illuminate the equatorial area of the target. Finally, for pulsed-power IFE systems, there may be no driver beams at all; the electrical energy is coupled directly to the target by the pressure of the magnetic field produced by the drive current.

The two principal methods of driving laser implosions are indirect drive and direct drive (see Figure 1.4). For ion accelerators, there is nearly a continuum between indirect drive and direct drive.

The three principal methods proposed to ignite the fuel are referred to as hot-spot ignition, shock ignition, and fast ignition. For indirect drive, there is some thermal inertia or heat capacity associated with the cavity surrounding the fuel capsule and with the ablator itself. It is more difficult to achieve the rapid rise in temperature and pressure with indirect drive because of the thermal inertia of the hohlraum. Shock ignition requires rapidly rising drive pressure at the end of the drive pulse. Consequently, shock ignition is usually associated with direct drive. Hot-spot ignition and fast ignition are the main ignition modes for indirect drive. All three modes of ignition necessarily ignite only a small fraction of the fuel. The thermonuclear burn then propagates into the bulk of the fuel.

Implosion Requirements

A number of conditions must be satisfied to produce ignition and reactor-scale gain.[2] These conditions are described in detail in Appendix A; in this section, the committee gives a brief overview.

---

[2] R. Betti, University of Rochester, "Tutorial on the Physics of Inertial Confinement Fusion for Energy Applications," Presentation to the committee on March 29, 2011.

Symmetry

Ideally, the final configuration of the imploded fuel should be nearly spherical. For laser-driven and heavy-ion-driven implosions, this requirement imposes conditions on the uniformity of the light, X-ray, or ion flux driving the target, and also on the initial uniformity of the target itself. For example, if the target is driven more strongly near the poles, the final imploded configuration might be shaped like a pancake. If the equator is driven more strongly, the imploded configuration might resemble a sausage. The greater the convergence ratio[3] of the target, the greater the precision required in direct drive—for example, in drive pressure or shell thickness. For most laser target designs, this convergence ratio lies between 20 and 40.

Sausagelike, pancakelike, dumbbell-like, or even doughnutlike asymmetries are low-order asymmetries in the sense that the wavelength of the departures from spherical symmetry is comparable to the size of the compressed fuel configuration. Energy imbalance among the beams is one possible type of error leading to low-order asymmetries; beam misalignment is another.

Fluid Instabilities

In addition to the low-order asymmetries, higher-order asymmetries are also important. Small perturbations on the surfaces of the fuel and ablator shell can grow as the shell is accelerated.

Unless the initial layer surfaces are very smooth (i.e., perturbations are smaller than about 20 nm), short-wavelength (wavelength comparable to shell thickness) perturbations can grow rapidly and destroy the compressing shell.

Mix

Similarly, near the end of the implosion, such instabilities can mix colder material into the spot that must be heated to ignition. If too much cold material is injected into the hot spot, ignition will not occur.

Density

Most of the fuel must be compressed to high density, approximately 1,000 to 4,000 times solid density. (In the case of hot-spot ignition, the central (gaseous) portion of the fuel is compressed to lower density.) Compression to such high densities demands that the fuel remain relatively cool during compression—technically,

---

[3] For hot-spot ignition, the convergence ratio is usually defined as the initial target radius divided by the final hot-spot radius.

very nearly Fermi-degenerate. Otherwise, too much energy is required to achieve the required density. This requirement in turn places stringent constraints on the pulse shape driving the target. The drive pressure must initially be relatively low (on the order of 1 Mbar); otherwise the initial shock wave that is created will heat the fuel to an unacceptable level. The pressure must then increase to produce a sequence of carefully timed shock waves to compress and ignite the fuel in the hot spot. Moreover, if the beam–target interaction produces too many energetic electrons or photons that can penetrate into the fuel and preheat it, efficient compression is not possible.

Fuel compression is related to an important quantity, the product of fuel density and fuel radius (ρr). This quantity is important for two reasons. The first is related to ignition. Ignition occurs when the rate of energy gain in the fuel exceeds the rate of energy loss. The igniting fuel gains energy as the fuel is shocked and compressed, but it must also gain energy by capturing its own burn products; specifically, in the case of deuterium-tritium fuel, it must capture the alpha particles that are produced. In this case, the ρr of the hot spot must exceed approximately 0.3 g/cm^2, the stopping range of an alpha particle in igniting fuel.[4] The second reason that ρr is an important quantity is because it determines the fraction of fuel that burns. This fraction is approximately given by $\rho r /(\rho r + 6)$, where ρr is given in g/cm^2. To achieve high target energy gain needed for laser inertial fusion energy (IFE), the ρr of the entire fuel, not just the hot spot, must be of the order of 3 g/cm^2. It is noteworthy that if one were to achieve such a ρr with uncompressed fuel, the fuel mass would be of the order of 1 kg. Heating 1 kg to 10 keV requires about 10^{12} J (~200 tons of high explosive equivalent) delivered to the fuel, and the resulting fusion yield would be 100 kton. These are perhaps the most important reasons why a small mass of fuel, typically 1 to 10 mg, must be compressed to high density.

Implosion Velocity

As noted above, ignition occurs when the rate of energy gain in the fuel exceeds the rate of energy loss. For hot-spot ignition, an implosion velocity on the order of 300 km/s is required to provide adequate self-heating of the fuel. It is fortunate that this velocity corresponds to a specific energy that is more than adequate to compress the fuel to the required density. However, since the ignition velocity exceeds the velocity needed for compression, it may be possible to improve target performance by separating the compression and ignition processes. This possibility is the reason for considering fast ignition and shock ignition.

---

[4] R. Betti, University of Rochester, "Tutorial on the Physics of Inertial Confinement Fusion for Energy Applications," Presentation to the committee on March 29, 2011.

Laser Targets, Direct and Indirect Drive

As discussed above, there are two principal ways to drive laser targets, direct drive and indirect drive. Both have advantages and disadvantages. Choosing between the two approaches has been, and remains, one of the most thoroughly (sometimes hotly) debated issues in inertial fusion. The choice is complicated because it involves not only target physics but also issues associated with target fabrication, reactor chamber geometry and wall protection, target injection, alignment tolerances, and target debris. Moreover, target performance depends on the wavelength and bandwidth of the laser light used to illuminate the target. Traditionally this dependence has coupled the choice of direct vs. indirect drive to the choice of laser, further complicating the scientific issues.

It is important that the laser–target interaction does not produce energetic photons or electrons that can preheat the fuel and prevent proper compression. A number of laser–plasma instabilities are known to produce preheat. The product of laser intensity (power per unit area) and wavelength squared is a measure of the importance of such instabilities. The instabilities are less important at lower intensities and shorter wavelengths. Consequently, as explained later in this chapter, solid-state lasers that typically produce light with a wavelength of 1 μm employ frequency doubling, tripling, or quadrupling to obtain wavelengths that are more compatible with target requirements. KrF lasers intrinsically produce light with a wavelength of 0.25 μm and do not require frequency multiplication. Even at shorter wavelengths, important concerns and uncertainties remain, especially because the targets required for inertial fusion power production must be larger than the targets that have been experimentally studied. Instabilities are expected to be worse in the larger plasma scale lengths associated with these larger targets.

The high efficiency of coupling laser energy to the imploding fuel is usually considered the most important advantage of direct drive. In the case of indirect drive, a substantial fraction of the laser energy must be used to heat the hohlraum wall. Typically less than half the laser energy is available as X-rays that actually heat the ablator. On the other hand, the calculated efficiency of X-ray ablation is usually somewhat higher than the efficiency of direct ablation, partially offsetting the hohlraum losses. Nevertheless, the higher coupling efficiency of direct drive is reflected in the target gain curves (target energy gain vs. laser energy) shown to the committee. Specifically, for hot-spot ignition, the calculated target gain for direct drive at the same drive energy is roughly 3 times higher, or, alternatively, 1.5 times higher at two-thirds of the drive energy. (Higher gain and lower driver energy lead to improved economics for IFE.) If shock ignition (described below) turns out to be feasible for direct drive but not indirect drive, the difference in gain between direct and indirect drive for a given driver energy will be more pronounced.

Another potential advantage of direct drive is the chemical simplicity of the target. Laser direct-drive targets usually contain little high-Z material. In contrast, indirect-drive targets require a hohlraum made of some high-Z material such as lead. For this reason the indirect-drive waste stream (from target debris) contains more mass and is chemically more complex than the direct-drive waste stream. This issue is discussed more fully in Chapter 3.

Indirect drive also has a number of advantages. For indirect drive, the beams do not impinge directly on the capsule but rather on the inside of the hohlraum wall (see Figure 1.4). The radiation produced at any point illuminates nearly half the surface area of the target. Moreover, the radiation that does not strike the target is absorbed and reemitted by the hohlraum wall. Thus, there is a significant smoothing effect associated with indirect drive. Consequently, beam uniformity, beam energy balance, and beam alignment requirements are less stringent than they are for direct drive. For example, for direct drive, a typical beam alignment tolerance might be 20 μm. The NIF baseline indirect-drive target, however, can tolerate a beam misalignment of about 80 μm. Furthermore, although the hohlraum complicates the waste stream from the target, it also provides thermal and mechanical protection for the target as it is injected into the hot chamber. This protection enables the use of chamber wall protection schemes (e.g., gas protection) that are not available to direct drive; for instance, gas in the chamber produces unacceptable heating of bare, direct-drive targets. Moreover, the smoothing effects of the hohlraum allow greater flexibility in beam geometry (chamber design) than is the case for direct drive. Specifically, polar illumination is suitable for indirect drive. It is likely suitable for direct drive as well, but for direct drive it degrades performance relative to spherical drive.

A final advantage of indirect drive is not a technical advantage at all, but rather a programmatic advantage. Much of the capsule physics of indirect drive is nearly independent of the driver. Therefore significant amounts of the information learned on laser indirect-drive experiments carry over to indirect drive for ion-driven targets.

As for interactions with the chamber wall, direct-drive targets and indirect-drive targets have very different output spectra in terms of the fraction of energy in exhaust ions compared to the fraction of energy in X-rays. Specifically, for indirect drive a substantial fraction of the ion energy is converted to X-rays when the ions strike the hohlraum material. Partly because of the difference in spectra, different wall protection schemes are usually adopted for the two target options. For example, magnetic deflection of ions is an option that is being considered for direct drive while gas or liquid wall protection to absorb X-rays is usually favored for indirect drive. The issues of output spectra, target debris, chamber options, and target fabrication costs are discussed more fully in Chapter 3.

The NIF houses the world's largest operating laser.[5] The NIF team has selected indirect drive with hot-spot ignition and polar illumination for its first ignition experiments. Without modification, the NIF could also be used to study some aspects of direct drive such as the behavior of laser beams in plasmas having large scale lengths. With modifications to improve beam smoothness, NIF is also able to study polar direct drive with and without shock ignition.[6] Such modifications are estimated to take 4 or more years to complete and cost $50 million to $60 million (including a 25 percent contingency added by this committee; see Chapter 4).[7]

In summary, both direct drive and indirect drive have advantages. The current uncertainties in target physics are too large to determine which approach is best, particularly when one includes all the related issues associated with chambers, target fabrication and injection, wavelength dependence, and so on. This conclusion leads to Recommendation 2-1, below.

Laser-Driven Fast Ignition

In laser-driven fast ignition the target is compressed to high density with a low implosion velocity and then ignited by a short, high-energy pulse of electrons or ions induced by a very short (a few picoseconds) high-power laser pulse.[8] Fast ignition has two potential advantages over conventional hot-spot ignition: higher gain, because the target does not need to be compressed as much, and relaxed symmetry requirements, because ignition does not depend on uniform compression to very high densities. The fast-ignition concept for inertial confinement fusion (ICF) was proposed with the emergence of ultrahigh-intensity, ultrashort pulse lasers using the chirped-pulse-amplification (CPA) technique. The target compression can be done by a traditional driver: direct-drive by lasers or ion beams; or indirect drive from X-rays using a hohlraum driven by nanosecond lasers, ion beams, or a Z-pinch or magnetically imploded target. The ignition is initiated by a converting a short, high-intensity laser pulse (the so-called "ignitor pulse") into an intense electron or ion beam that will efficiently couple its energy to the compressed fuel.

A number of different schemes for coupling a high-energy, short-pulse laser to a compressed core have been examined. The "hole-boring" scheme involves

---

[5] E.I. Moses, 2011, The National Ignition Facility and the promise of inertial fusion energy, *Fusion Science and Technology* 60: 11-16.

[6] J. Quintenz, NNSA, and M. Dunne, LLNL, Two presentations to the committee on February 22, 2012 (see Appendix C).

[7] "Polar Drive Ignition Campaign Conceptual Design," TR-553311, submitted to NNSA in April 2012 by the Lawrence Livermore National Laboratory (LLNL) and revised and submitted to NNSA by the Laboratory for Laser Energetics (LLE) in September 2012.

[8] R. Betti, University of Rochester, "Tutorial on the Physics of Inertial Confinement Fusion for Energy Applications," Presentation to the committee on March 29, 2011.

two short-pulse laser beams, one having a ~100 ps duration to create a channel in the coronal plasma surrounding the imploded dense fuel, through which the high-intensity laser pulse that generates the energetic electrons or ion beams would propagate.[9] An alternative design uses a hollow gold cone inserted in the spherical shell,[10] as illustrated in Figure 2.1.

In this scheme, the fuel implosion produces dense plasma at the tip of the cone, while the hollow cone makes it possible for the short-pulse-ignition laser to be transported inside the cone without having to propagate through the coronal plasma and enables the generation of hot electrons at its tip, very close to the dense plasma. A variant cone concept uses a thin foil to generate a proton plasma jet with multi-MeV proton energies. The protons deliver the energy to the ignition hot spot, with the loss of efficiency in the conversion of hot electrons into energetic protons balanced by the ability to focus the protons to a small spot.[11]

As is the case for hot-spot ignition, the minimum areal density for ignition at the core ($\rho r \sim 0.3$ g/cm^2 at 10 keV) is set by the 3.5-MeV alpha particle range in deuterium-tritium (DT) and the hot-spot disassembly time. This must be matched by the electron energy deposition range. This occurs for electron energy in the ~1 to 3 MeV range. The minimum ignition energy, E_{ig}, is independent of target size and scales only with the density of the target; the greater the mass density, the less the beam energy required for ignition (about 20 kJ of collimated electron/ion beam energy is required for a ~300 g/cm^3 fuel assembly).[12]

The optimum compressed-fuel configuration for fast ignition is an approximately uniform-density spherical assembly of high-density DT fuel without a central hot spot. High densities can be achieved by imploding thick cryogenic DT shells with a low-implosion velocity and low entropy. Such massive cold shells produce a large and dense DT fuel assembly, leading to high gains and large burn-up fractions.

Experimental investigations of the fast-ignition concept are challenging and involve extremely high-energy-density physics: ultraintense lasers (>10^{19} W cm^{-2}); pressures in excess of 1 Gbar; magnetic fields in excess of 100 MG; and electric fields in excess of 10^{12} V/m. Addressing the sheer complexity and scale of the problem inherently requires the high-energy and high-power laser facilities that are now becoming available (OMEGA Extended Performance and NIF's Advanced

---

[9] M. Tabak, J. Hammer, M.E. Gilinsky, et al., 1994, Ignition and high gain with ultrapowerful lasers, *Physics of Plasmas* 1: 1626.

[10] R. Kodama, P.A. Norreys, K. Mima, et al., 2001, Fast heating of ultrahigh-density plasma as a step towards laser fusion ignition, *Nature* 412: 798.

[11] M.H. Key, 2007, Status of and prospects for the fast ignition inertial fusion concept, *Physics of Plasmas* 14: 5.

[12] R.R. Freeman, C. Anderson, J.M. Hill, J. King, R. Snavely, S. Hatchett, M. Key, J. Koch, A. MacKinnon, R. Stephens, and T. Cowan, 2003, High-intensity lasers and controlled fusion, *European Physics Journal D* 26: 73-77.

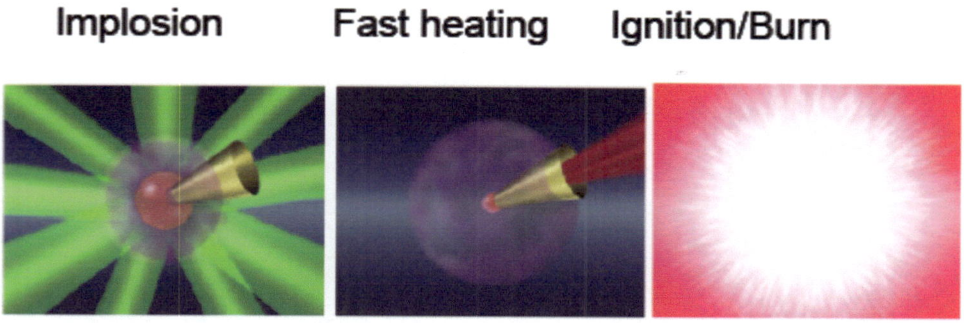

FIGURE 2.1 In this fast ignition approach, a hollow gold cone inserted in the spherical shell is used to couple energy to the compressed core. SOURCE: H. Azechi, Osaka University, "Inertial Fusion Energy: Activities and Plans in Japan," Presentation to the committee on June 15, 2011.

Radiographic Capability, among others) as well as the most advanced theory and computer simulation capability available.

Laser-Driven Shock Ignition

As in fast ignition, shock ignition separates the compression of the thermonuclear fuel from the ignition trigger. The ignition process is initiated by a spherically convergent strong shock (the "ignitor shock") launched at the end of the compression pulse. This late shock collides with the return shock driven by the rising pressure inside the central hot spot and enhances the hot-spot pressure.[13] Since the ignitor shock is launched when the imploding shell is still cold, the shock propagation occurs through a strongly coupled, dense plasma. If timed correctly, the shock-induced pressure enhancement triggers the ignition of the central hot spot. In laser direct-drive shock ignition, the capsule is a thick wetted-foam shell[14,15] driven at a relatively low implosion velocity of ~250 km/s. The compression pulse consists of a shaped laser pulse designed to implode the capsule with low entropy to achieve high volumetric and areal densities. The fuel mass is typically greater for shock ignition than for hot-spot ignition. The large mass of fuel leads to high fusion-energy yields and the low entropy leads to high areal densities and large burn-up fractions. These conditions lead to high predicted gain. The ignitor shock

---

[13] R. Betti, C.D. Zhou, K.S. Anderson, L.J. Perkins, W. Theobald, and A.A. Solodov, 2007, Shock ignition of thermonuclear fuel at high areal density, *Physical Review Letters* 98: 155001.

[14] Ibid.

[15] J. Sethian and S. Obenschain, Naval Research Laboratory, "Krypton Fluoride Laser Driven Inertial Fusion," Presentation to the committee on January 29, 2011.

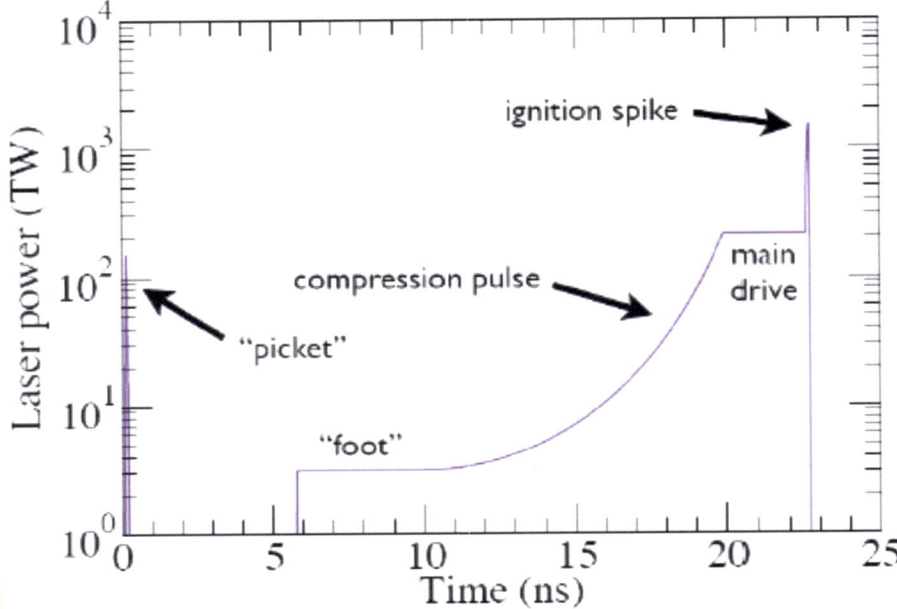

FIGURE 2.2 Shock ignition power input. SOURCE: J. Sethian and S. Obenschain, Naval Research Laboratory, "Krypton Fluoride Laser-Driven Inertial Fusion," Presentation to the committee on January 29, 2011.

is required because at low velocities the central hot spot is too cold to reach the ignition condition with the conventional ICF approach. The ignitor shock can be launched by a spike in the laser intensity on target or by particle beams incident on the target surface (see Figure 2.2).

Recent numerical simulations suggest that it may be possible to achieve gains exceeding 100 at laser energies smaller than 500 kJ.[16] Although the intensity of the final shock ignition pulse exceeds the threshold for laser–plasma instabilities, there are grounds to believe that target preheat by fast electrons may not be a problem.[17]

Laser Beam–Target Interaction

In order to achieve any of the conditions needed for ignition and thermonuclear burn, it is essential that the beams interact properly with the target. For

---

[16] A.J. Schmitt, J.W. Bates, S.P. Obenschain, S.T. Zalasek, and D.E. Fyfe, 2010, Shock ignition target design for inertial fusion energy, *Physics of Plasmas* 17: 042701.

[17] Ibid.

example, if too large a fraction of the beam energy is reflected or refracted away from the target, it is not possible to achieve high energy gain. Also, as noted above, the beam–target interaction must not produce a sufficient number of energetic electrons or photons to preheat the fuel so that it cannot be adequately compressed. For indirect drive, the beam energy must efficiently convert into X-rays, and for direct drive, the ablation process must efficiently drive the implosion. Despite extensive theoretical and experimental work, beam-target interactions are still not fully understood. The beam-target interaction for ion beams will be discussed in a later section. For laser beams, effects such as laser–plasma instabilities depend on the size of the plasma. While there is considerable experimental information at scale sizes that are too small to achieve ignition and burn, these instabilities are an important concern for both direct drive and indirect drive for fusion-scale targets, especially because the available experimental data are limited. Furthermore, the instabilities become more deleterious with increasing wavelength and increasing laser intensity. The scaling with wavelength is the reason that current target experiments are usually performed with frequency-tripled 351 nm light from solid-state lasers or the 248 nm ultraviolet light from KrF lasers. The intensity scaling means that laser–plasma instabilities are greater during the brief shock-ignition pulse than during hot-spot ignition, although hot-spot ignition may be more vulnerable to the hot electrons produced by laser–plasma instabilities over the long drive pulse. OMEGA, Nike, and the NIF are valuable national assets that are continuing to elucidate the unknown features of laser–plasma interactions.

Status of Laser-Driven Target Implosion Research

The NIF laser, commissioned in March 2009, is a unique facility for exploring IFE physics and validating target design and performance. It is the only facility that may be able to demonstrate laser-driven ignition during the next several years. It can deliver up to ~1.8 MJ of UV (351 nm) energy with 30-ps timing precision. The NIF laser has met a 95 percent availability level for requested shots, and more than 300 shots were commissioned through 2012. Critical ignition physics studies took place during the National Ignition Campaign (NIC) program, which concluded on September 30, 2012. The goal of this program was to achieve ignition, to commission targets, and to understand the physics necessary for successful, reliable ignition. Recent target shots have led to improved symmetry and a measured yield of 5-9 × 10^{14} neutrons at 1.4-1.6 MJ drive energy. To put this in perspective, alpha particle heating of dense fuel surrounding the hot spot is confirmed at a yield of ~10^{16} neutrons and breakeven ignition at ~5.6 × 10^{17} neutrons on a threshold curve calculated to be very steep.[18] The NIC made progress in approaching the sphericity, compression, and velocity needed for ignition. However, the NIC experiments produced a number of surprising results, particularly a lower-than-expected implosion velocity. There are also still uncertainties associated with low-mode asymmetries of the dense fuel and mix.

In its conclusion, the Target Panel (see Appendix H for the summary of its report) says that "based on its analysis of the gaps in current understanding of target physics and the remaining disparities between simulations and experimental results ... ignition using laser indirect drive is not likely in the next several years."[19] In the same place, it also states that "resolving the present issues and addressing any new challenges that might arise are likely to push the timetable for ignition to 2013-2014 or beyond." The panel goes on to also conclude as follows:

- If ignition is achieved with indirect drive at NIF, then an energy gain of 50-100 should be possible at a future facility. How high the gain at NIF could be will be better understood by follow-on experiments once ignition is demonstrated. At this writing, there are too many unknowns to project a potential gain. (Conclusion 4-3)
- Achieving ignition will validate assumptions underlying theoretical predictions and simulations. This may allow a better appreciation of the sensitivities to parameters important to ignition. (Conclusion 4-3)
- The NIF has the potential to support the development and further validation of physics and engineering models relevant to several IFE concepts, from indirect-drive hohlraum designs to polar direct-drive ICF and shock ignition. (Overarching Conclusion 1)
- The NIF will also be helpful in evaluating indirectly driven, heavy-ion targets. It will be less helpful in gathering information relevant to current Z-pinch, heavy-ion direct drive, and heavy-ion advanced target concepts.

As noted above, the NIC was completed on September 30, 2012. With input from the ICF laboratories, the National Nuclear Security Administration (NNSA) produced a report that put forward a "Plan B" experimental program for FY2013 and beyond.[20] These issues and tentative plans were discussed in presentations to the committee.[21]

Conclusion 2-1: There has been good technical progress during the past year in the ignition campaign carried out on the National Ignition Facility.

---

[18] E.I. Moses, 2011, The National Ignition Facility and the promise of inertial fusion energy, *Fusion Science and Technology* 60: 11-16.

[19] NRC, 2013, *Assessment of Inertial Confinement Fusion Targets*, Washington, D.C.: The National Academies Press, released as a prepublication (Target Panel Report).

[20] National Nuclear Security Administration, 2012, *NNSA's Path Forward to Achieving Ignition in the Inertial Confinement Fusion Program: Report to Congress*, December.

[21] J. Quintenz, NNSA, and M. Dunne, LLNL, Two presentations to the committee on February 22, 2012 (see Appendix C).

Nevertheless, ignition has been more difficult than anticipated and was not achieved in the National Ignition Campaign, which ended on September 30, 2012. The results of experiments to date are not fully understood. It will likely take significantly more than a year to gain a full understanding of the discrepancies between theory and experiment and to make modifications needed to optimize target performance.

The NIF is currently a unique tool for addressing these issues, some of which could be addressed with NIF in its present configuration. Others may require modifications such as improvements in beam smoothness or, ultimately, even a different illumination geometry.

Laser–plasma instabilities (LPI) are present in current NIF indirect-drive experiments as well as in the most energetic spherical direct drive (SDD) experiments performed on OMEGA. Robust, high-gain, laser inertial fusion target design must address and contain the effects of these nonlinear processes, which have an intensity threshold behavior that in principle makes modeling extrapolation from low gain to high gain problematic. Both OMEGA (glass laser) and Nike (KrF laser) can test different ablator materials with respect to laser–plasma instabilities. Following the recent results from OMEGA experiments,[22] ablators with moderate atomic number (from carbon to silicon) greatly reduce LPI while preserving good hydrodynamic properties. OMEGA and Nike can also compare the acceleration of flat foils at the different wavelengths of 351 nm (OMEGA) and 249 nm (Nike), with different bandwidths or beam smoothing, to determine whether there is a significant advantage to using shorter-wavelength, higher-bandwidth KrF illumination for direct drive. Options to continue the work are discussed in the subsection Laser Drivers, below.

Recommendation 2-1: The target physics programs on the NIF, Nike, OMEGA, and Z should receive continued high priority. The program on NIF should be expanded to include direct drive and alternate modes of ignition. It should aim for ignition with moderate gain and comprehensive scientific understanding leading to codes with predictive capabilities for a broad range of IFE targets.

Ion Beam Targets

In many respects, ion beam targets are similar to the laser targets that have just been discussed. Ion range (penetration depth) is roughly the analog of laser

---

[22] V. Smalyuk, R. Betti, J.A. Delettrez, V. Yu, et al., 2010, Implosion experiments using glass ablators for direct-drive inertial confinement fusion, *Physical Review Letters* 104: 165002.

wavelength. Ion range is a function of ion mass and ion kinetic energy. The range decreases with increasing mass and increases with increasing kinetic energy. Light ions (e.g., Li) have the appropriate range to drive targets at a kinetic energy on the order of 30 MeV. Heavier ions such as Cs or Pb have the appropriate range at energies in the multi-GeV range. It is usually easier to focus ions at higher kinetic energy and higher mass, so most of the emphasis is currently on heavy-ion fusion as opposed to light-ion fusion. Nevertheless, the comments in this section apply to both.

For ion indirect drive, the fuel capsule (the ablator and fuel) is essentially the same as the fuel capsule for laser indirect drive. The primary difference lies in the physics of the beam–target interaction and conversion of beam energy into radiation. Thus, experience with laser indirect drive on the NIF will put to rest many of the issues associated with ion indirect drive.[23] In this regard, it important to note that target simulations for both driver options are performed using the same computer codes. From a fuel-capsule standpoint, the status and issues are the same as those discussed above for laser indirect drive. The principal new questions are these:

- Can one correctly predict the range of intense ion beams in hot matter?
- Are there processes that can produce unacceptable levels of preheat?
- What is the efficiency of converting beam energy into radiation?

Ion range has been studied for nearly a century. The theory is relatively straightforward, and the agreement between theory and experiment is good for low-intensity ion beams in cold matter. In particular, numerous ion deposition experiments have been performed in the kinetic energy range of interest for both light-ion and heavy-ion fusion. The range of intense ion beams in hot matter is the question. Some experiments have been performed in preheated plasmas to simulate the conditions appropriate for inertial fusion, and light-ion beams have been used to heat material to 58 eV, at temperatures within a factor of ~3 of that needed for inertial fusion.[24] The theoretical uncertainties in ion range in hot matter appear to have little relevance for indirectly driven targets, since the beam energy, the target material(s), and the wall thickness can be adjusted when the details of ion–beam–matter interaction are actually measured.

There have also been extensive theoretical and numerical searches for processes that might produce unacceptable preheat.[25] No such processes have been found.

---

[23] J.D. Lindl, P. Amendt, R.L. Berger, S.G. Glendinning, et al., 2004, The physics basis for ignition using indirect-drive targets on the National Ignition Facility, *Physics of Plasmas* 11(2): 339.

[24] Ibid.

[25] D.W. Hewett, W.L. Kruer, and R.O. Bangerter, 1991, Corona plasma instabilities in heavy-ion fusion targets, *Nuclear Fusion* 31(3): 431 and references therein.

Also, numerical simulations predict high conversion efficiency of ion-beam energy into radiation.

In summary, calculations and limited experimental information are promising for ion-beam indirect drive. Numerical simulations predict gains as high as 130 at 3 MJ, but experiments with more intense beams are required to augment the information on indirect-drive target performance being produced at the NIF.

For lasers, it is appropriate to make a sharp distinction between direct drive and indirect drive. For ion beams, the distinction is not as sharp. There are targets that are fully directly driven or fully indirectly driven, but there are also targets that lie between the two extremes. Calculations indicate that the targets at the direct end of the spectrum can produce high gain at low driver energy.[26] Unfortunately, the ion range needed for pure direct drive is sufficiently small that it has proved very difficult to design an accelerator that can meet the focusing requirements. This situation has led to the study of targets that are similar to directly driven targets except that the outer shell of the target, outside the ablator, is made of a dense, high-Z material. Early in time, the pressure to drive the implosion is almost completely generated by direct ion deposition, i.e., by direct drive. Later in the pulse, radiation becomes an important energy transport mechanism and the dense shell acts like a hohlraum. Calculations indicate that these targets can also produce high gain at low driver energy. Moreover, the gain is relatively insensitive to ion range, and the ion range is comparable to that required by indirect drive. These "mixed" targets are often referred to as directly driven targets, although the physics of the implosion and issues of stability are very different from those used in laser direct drive.

Currently there are ongoing numerical simulations involving direct drive with hot-spot ignition and shock ignition. Both spherical and polar illumination geometries are being considered. As is the case for lasers, the predicted target gain is higher for direct drive than for indirect drive. Unfortunately, there is no experimental information on ion direct drive.

Ion-Driven Fast Ignition

The earliest targets for heavy-ion fusion, described in the mid-1970s, were based on fast ignition using intense ion beams.[27] Imploding the fuel using ion beams and igniting it with a laser is another option. Current research favors the original approach, which uses ion beams for both processes. In principle, one should be able to achieve high gain from such targets. Also, the ignition physics appears to be more straightforward than laser fast-ignition physics, but the ion kinetic energy

---

[26] B.G. Logan, LBNL, "Heavy Ion Fusion," Presentation to the committee on January 29, 2011.

[27] A.W. Maschke, 1975, Relativistic ions for fusion applications, Proceedings of the 1975 Particle Accelerator Conference, Washington, D.C., *IEEE Transactions on Nuclear Science*, NS-22(3): 1825.

required to obtain the required small focal spots is an order of magnitude or more larger than the kinetic energy required for direct drive or indirect drive. Although the ignition physics appears to be straightforward, some important parts of this physics have not yet been incorporated into the codes used for numerical simulation. Furthermore, there are important uncertainties in focusing physics, target physics, and accelerator design that have not been adequately addressed. If these uncertainties can be resolved favorably using theory and simulation, there is still a programmatic issue. The accelerator needed to drive fast ignition targets is not the accelerator needed to drive the other types of targets. In other words, to obtain definitive experimental information on this option, one would have to build a unique accelerator with a far shorter pulse length. The challenges for this approach are to address the uncertainties, establish its superiority over other approaches, and develop a strong enough case to build a unique accelerator.

It is noteworthy that both U.S. and foreign heavy-ion fusion programs are studying targets based on ion fast ignition. The U.S. version of such targets is referred to as the X-target (see Figure 2-6 in the Target Panel report). The X-target design has evolved rapidly during the last year but has not been fully evaluated.

Pulsed-Power Targets

Historically, both indirect drive and ion- and electron-driven direct drive have been studied for pulsed-power inertial fusion. Many of the considerations discussed above for laser and heavy-ion targets also apply to these classes of pulsed-power targets. Magnetic implosion offers the possibility of significantly higher implosion efficiency than the other approaches, and it is currently the favored option. The targets being considered for Magnetized Liner Inertial Fusion at present are beryllium (conducting) cylinders that contain the fusion fuel at high pressure. As the magnetically driven implosion of the cylinder is initiated, a laser preionizes and preheats the gaseous fuel, which is then compressed and heated to ignition by the imploding metal cylinder in less than 100 ns (see Figure 2.3). The codes used to design these targets have not yet been experimentally validated.[28]

In the case of Magnetized Target Fusion (MTF), a field-reversed-configuration plasma is compressed by an imploding metal cylinder on a timescale of a few microseconds.[29]

---

[28] M. Cuneo et al., Sandia National Laboratories, "Pulsed Power IFE: Background, Phased R&D and Roadmap," Presentation to the committee on April 1, 2011.

[29] G. Wurden and I. Lindemuth, "Magneto-Inertial Fusion (Magnetized Target Fusion)," Presentation to the committee on March 31, 2011.

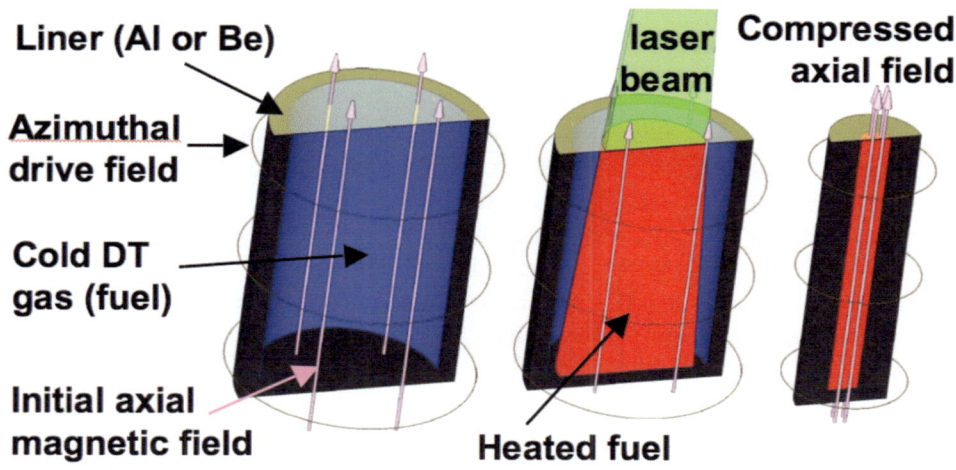

FIGURE 2.3 The magnetized liner fusion target. SOURCE: M. Cuneo, Sandia National Laboratories, Presentation to the committee on April 1, 2011.

DRIVER OPTIONS FOR INERTIAL CONFINEMENT FUSION

This section provides a description of each driver type being considered for IFE. Each driver description begins with background and status of the driver technical application and then goes on to the scientific challenges and future research and development priorities, including a description of the path forward in the near, medium, and long term for each driver type.

As noted in the preceding section, the technical approaches to achieving inertial fusion energy include three kinds of drivers: lasers, heavy-ion accelerators, and electrical pulsed-power systems. As discussed below, good progress has been made in developing the repetitively pulsed systems required for fusion energy. Nevertheless, for all types of drivers, there remain substantial challenges in developing systems that would have the quality, reliability, maintainability, and availability to provide a number of shots that, depending on the driver, range from 3×10^6 to 4×10^8 per year. For each technological approach, the committee identifies a series of critical R&D objectives that must be met for that approach to be viable. If these objectives cannot be met, then other approaches will need to be considered.

Laser Drivers

Two types of laser drivers have been considered as possible candidates for IFE: the solid-state laser and the krypton fluoride (KrF) gas laser. The first part of this section describes progress in solid-state laser technology. The second part describes the background and progress in KrF ultraviolet gas laser for fusion-driver applications.

All lasers require a gain medium, a pump source, and an optical resonator system to shape and extract the laser power. Since the demonstration of the lamp-pumped, ruby laser in 1960, enormous progress has been made in the gain media, pumping sources, operating efficiency, and average power of lasers. A recently published handbook provides an overview of the status of high-power lasers, including chapters on the NIF laser, the KrF laser, and on high-power diode arrays for pumping high-average-power, solid-state lasers.[30]

Projected Target Gains

Ignition and gain with indirect drive are presently being pursued in the NIF, following decades of research on earlier laser systems such as Nova.[31] Computations at LLNL suggest that in a power plant, reactor-scale target gains of ≥60 might be attainable with optimized indirect-drive targets driven by 2 MJ of 3ω[32] light.[33]

Direct-drive targets are also being considered. Their designs evolved from work at the University of Rochester's LLE and the Naval Research Laboratory (NRL) during the 25 years from 1985 to 2010, taking advantage of the new smoothing techniques and tailored adiabats. In one-dimensional calculations, a reactor-scale target gain of 150 with only 400 kJ input has been projected when a 248-nm KrF wavelength is used with shock ignition; the calculated target gain vs. laser drive energy is shown in Figure 2.4.

Diode-Pumped Solid-State Lasers

Background and Status

Early solid-state lasers were pumped by spectrally broad flashlamps, from which only a small fraction of light was absorbed by the laser ions, leading to operating efficiencies in the range of 1-2 percent. The trend in commercial lasers is to replace lamp-pumped, solid-state lasers with diode-pumped, solid-state

---

[30] H. Injeyan and G.D. Goodno, 2011, *High-Power Laser Handbook*, New York, N.Y.: McGraw Hill.
[31] Nova is the 100 kJ, flashlamp-pumped laser that preceded the NIF at Lawrence Livermore National Laboratory.
[32] That is, three times the fundamental frequency of the laser, or 351 nm wavelength.
[33] M. Dunne, LLNL, "Update on NIF, NIC and LIFE," Presentation to the committee on February 22, 2012.

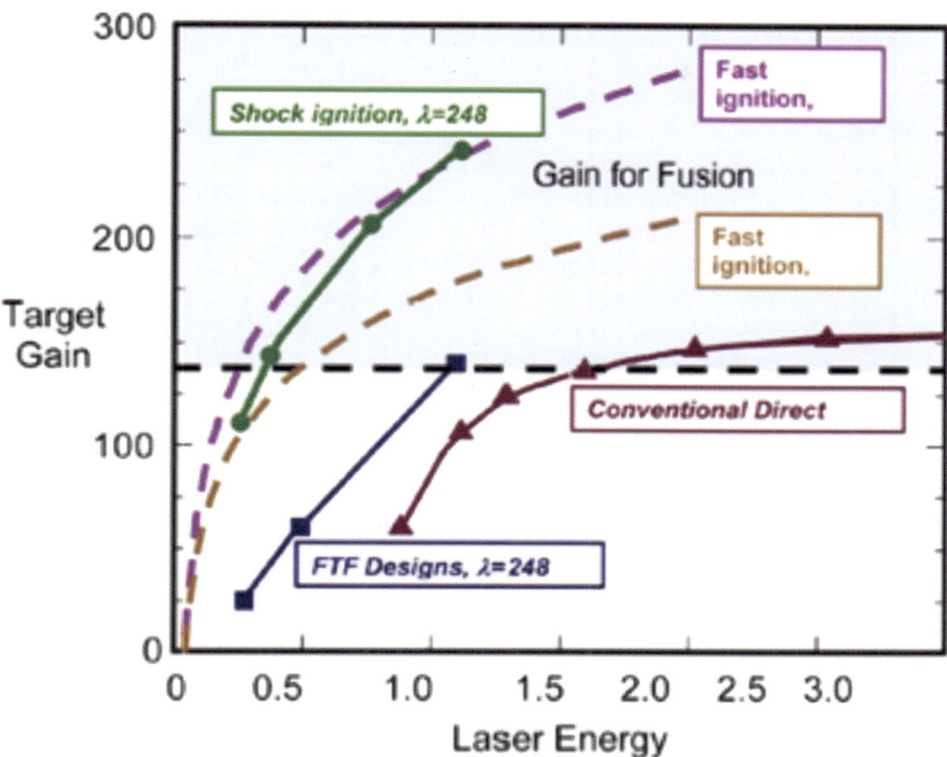

FIGURE 2.4 Target gain curves from one-dimensional simulations of various high-performance direct-drive target designs. The horizontal axis is in megajoules. The shaded region shows sufficient target gain for the power plant with KrF laser drive (G = 140). A gain G = 60 is shown as sufficient for a diode-pumped, solid-state laser (DPSSL) drive. Triangles are the calculated gain for a conservative conventional direct drive target, for either KrF or DPSSL (300 km/s implosion velocity). Squares are the Fusion Test Facility designs for KrF (λ = 248 nm) and higher ablation pressure implosion velocity of 350-450 m/s. Circles are for shock-ignition targets for KrF: soft conventional compression (<300 km/s) and then spike to shock heat to ignition. Dashed lines are fast ignition scaling for KrF (248 nm) and DPSSL (351 nm). Both fast ignition and shock ignition calculated gain curves are considered to be optimistic because so little is known about implementation. SOURCE: Adapted from J. Sethian, D.G. Colombant, J.L. Giuliani, et al., 2010, The science and technologies for fusion energy with lasers and direct-drive targets, *IEEE Transactions on Plasma Science* 38: 690.

lasers (DPSSLs) to improve operational efficiency and reliability for demanding 24/7 industrial applications.

An example solid-state laser consists of a diode laser tuned to 808 nm to match the absorption line of the neodymium (Nd) ion doped into a yttrium–aluminum–garnet (YAG) crystal. A lens focuses the diode output into the Nd:YAG crystal, and a resonator around the Nd:YAG crystal tuned to 1064 nm forms the oscillator.[34] To obtain higher power, the design is extended to the "master oscillator, power amplifier" configuration, where the low-power, well-controlled laser oscillator output is amplified by a power amplifier, as the name suggests. Today, solid-state lasers are commercially available with power levels ranging from ~1 W to 10 kW, and they operate with very high reliability to support manufacturing processes.

The scale of the laser energy required for an indirect-drive or direct-drive IFE power plant is likely to be comparable to the NIF laser—i.e., ~2 MJ per pulse in the ultraviolet but operated at 5 to 15 pulses per second repetition rate. Although a DPSSL driver can be used to drive either direct-drive or indirect-drive targets, this section describes a DPSSL-driven IFE power plant based on indirect drive because that approach is more mature and has been studied in the NIF-driven target experiments in depth. A KrF laser direct-drive approach is also discussed below. If direct drive proves to offer lower thresholds for ignition, as predicted by theory but not confirmed by experiments to date, then the DPSSL laser can be engineered to drive polar- or spherical-direct-drive targets.[35] For simplicity, in the remainder of the DPSSL section the term "laser" or "solid-state laser" will be used to mean "diode-pumped solid-state laser."

While the NIF laser was designed for single-shot operation for target physics and ignition studies, an IFE laser driver must operate at 5 to 15 shots per second for extended periods of time at high efficiency. As such, an IFE solid-state laser driver cannot be flashlamp-pumped, as is the NIF laser. For example, one proposed laser-driven, IFE power plant design, the laser inertial fusion energy (LIFE) design,[36] proposes to use DPSSLs and a modular architecture approach, as illustrated in Figure 2.5.

Laser system designs, based on extensive experimental measurements, show that advanced phosphate glass (APG) can operate at a 10-20 Hz repetition rate when diode-laser pumped at a safety margin of one-third the stress fracture limit.[37] Improvements in diode laser efficiency, diode laser-array irradiance, and coupling efficiency have allowed the projected electrical efficiency of solid-state IFE drivers to increase from 8.5 percent in 1996 to about 15 percent wall-plug efficiency (cooling taken into account) in the UV in a present-day energy-storage laser design.[38]

---

[34] R.L. Byer, 1988, Diode laser-pumped solid-state lasers, *Science* 239: 742-747.

[35] J. Quintenz, NNSA, "Status of the National Ignition Campaign & Plans Post-FY 2012," Presentation to the committee on February 22, 2012.

[36] T.M. Anklam, M. Dunne, W.R. Meier, S. Powers, and A.J. Simon, 2011, LIFE: The case for early commercialization of fusion energy, *Fusion Science and Technology* 60: 66-71; see also T. Anklam, LLNL, "LIFE Economics and Delivery Pathway," Presentation to the committee on January 29, 2011.

[37] A. Bayramian, S. Aceves, T. Anklam, et al., 2011, Compact, efficient laser systems required for laser inertial fusion energy, *Fusion Science and Technology* 60: 28-48.

[38] Ibid.

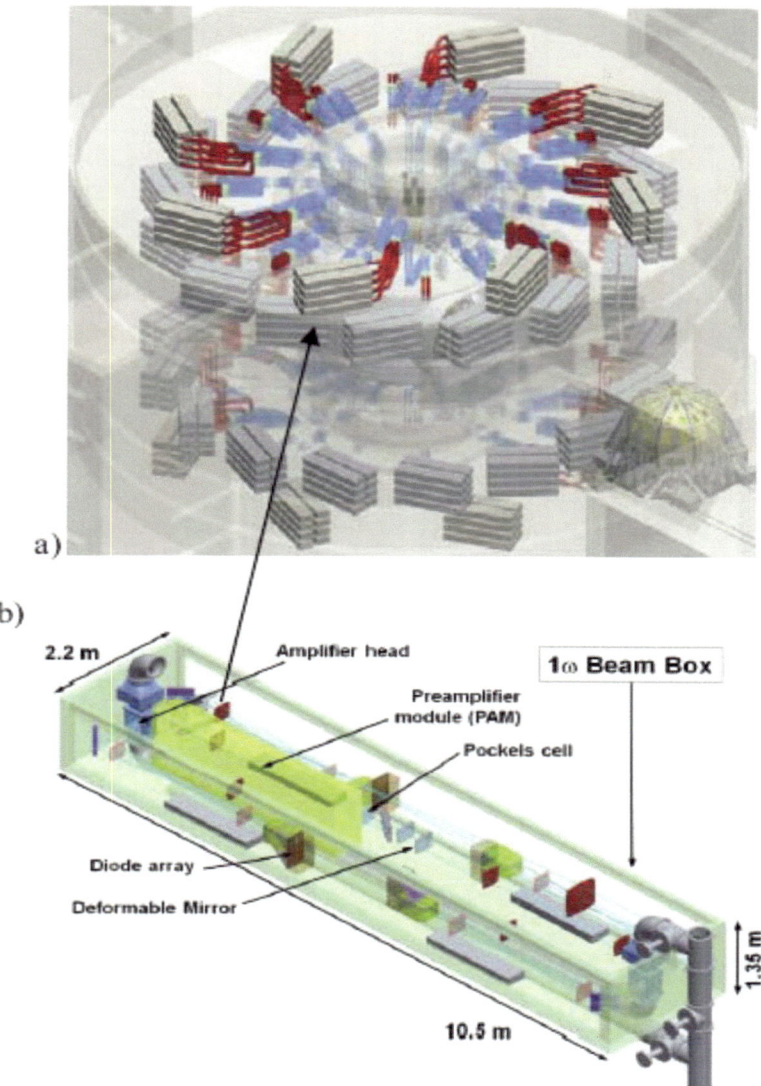

FIGURE 2.5 (a) Isometric view of a proposed laser-driven IFE power plant showing compact beam architecture composed of 384 lasers. (b) Isometric expanded view showing the contents of one ~100 kW solid-state laser in a beam box. SOURCE: J. Latkowski, LLNL, private communication to the committee, December 23, 2011.

As an example of average power and efficiency, a continuous-wave, diode-laser-pumped Nd:YAG laser, with more efficient power extraction than the pulsed laser for IFE, demonstrated greater than 19 percent wall plug efficiency in 2009 in a near-diffraction-limited beam at a 105 kW average power.[39]

The modular architecture provides flexibility in laser operation. For example, the laser can be configured to generate high-intensity green (frequency doubled) light at 532 nm. Green light often is associated with greater laser–plasma interaction (LPI) but offers the potential to assemble larger targets for higher gain. Further, the laser can generate output in the deep UV (4ω) at 263 nm for plasma studies or direct-drive studies. Recent work demonstrated near-room-temperature frequency doubling in a deuterated potassium dihydrogen phosphate (KDP) nonlinear crystal with 79 percent efficiency from a green Nd:glass laser to the deep UV at 263 nm.[40] This was achieved in a single-shot second harmonic generation experiment of the green 526 nm to generate UV at 263 nm at an intensity of 1 GW/cm^2 from a 3 ns, 4 J green pulse.

According to presentations to the committee, the global market for solid-state lasers has increased more than 15 percent per year, a pace that has facilitated mass production of laser diodes in a very competitive market served by many suppliers.[41] Commercial markets have driven continuous improvements in the performance and efficiency of laser diodes for pumping solid-state lasers. The size and the growth of the commercial markets underpin the projection of cost and performance of diode laser arrays for pumping future IFE solid-state laser drivers. Of particular interest are the projected lifetimes of large diode laser arrays for pumping an IFE laser driver. Based on recent measurements, the operational lifetimes are projected to be greater than 13.5 billion shots, or greater than 100,000 hours at a 37 Hz repetition rate.[42]

The semiconductor diode laser array manufacturers prepared a white paper stating that they can meet the projected costs and performance requirements for diode laser arrays for pumping solid state lasers for IFE.[43] This white paper

---

[39] J. Marmo, H. Injeyan, H. Komine, S. McNaught, J. Machan, and J. Sollee, 2009, Joint high power solid state laser program advancements at Northrop Grumman, *SPIE Proceedings* 7195: 719507.

[40] S.T. Yang, T. Steven, M.A. Henesian, T.L. Weiland, et al., 2011, Noncritically phase-matched fourth harmonic generation of Nd:glass laser in partially deuterated KDP crystals, *Optics Letters* 36: 1824.

[41] A.J. Bayramian, S. Aceves, T. Anklam, et al., 2011, "Compact, efficient laser systems required for laser inertial fusion energy," *Fusion Science and Technology* 60: 28-48. and R. Deri, J. Geske, M. Kanskar, S. Patterson, G. Kim, Q. Hartmann, F. Leibreich, E. Deichsel, J. Ungar, P. Thiagarajan, R. Martinsen, P. Leisher, E. Stephens, J. Harrison, C. Ghosh, O. Rabot, A. Kohl, "Semiconductor Laser Diode Pumps for Inertial Fusion Energy Lasers," Lawrence Livermore National Laboratory, LLNL-TR-465931, January 2011.

[42] R. Feeler, J. Junghans, J. Remley, D. Schnurbusch, and E. Stephens, 2010, Reliability of high-power QCW arrays, *Proceedings of SPIE* 7583.

[43] R. Deri et al., op. cit.

estimates a cost reduction to 0.7 cents per watt of diode laser light for an *n*th-of-a-kind IFE plant to be possible.[44]

An estimate of the cost of diode laser arrays versus the production volume has been made by engineers in Japan.[45] The projected costs, based on past and current diode laser costs, are $0.03/peak-watt at production volume of 100 million bars per year. This cost estimate appears to be consistent with that made at LLNL in their projections of diode laser costs.[46]

Table 2.1 describes the proposed design for an IFE driver operating in the UV at 351 nm with 2.2 MJ total energy and comprised of 384 lasers in a box. The top-level IFE laser driver system requirements are 2.2 MJ in the UV (351 nm) operating at 16 Hz repetition rate for an average laser power of 35 MW at 18 percent electrical efficiency (equivalent to 15 percent wall-plug efficiency) in the UV.

Details of the proposed solid-state IFE driver based on neodymium-doped APG are provided in a recent publication.[47] A single laser in a box module of the laser driver would operate at 130 kW (IR)/91 kW (UV) average power and 8.1 kJ (IR)/5.7 kJ (UV) output pulse energy at 16 Hz repetition rate. The aperture size is 25 × 25 cm and the operating UV wall-plug efficiency is 15 percent. The laser design would use a series of well-known features such as polarization rotation for birefringence compensation, flowing helium gas for cooling of the 20 graded-doped, 1-cm-thick APG glass gain elements in each of the two gain modules, and polarization combining of the diode laser pump arrays to double pump irradiance. The projected 75 percent harmonic conversion efficiency to the UV is obtained by optimizing harmonic conversion in separate channels for the foot and the peak of the laser pulse shape. Finally, the proposed modular architecture for the laser has a built-in 15 percent operating margin, such that the fusion plant could continue to operate even with the shutdown of a beam line for replacement or repair. The proposed laser-in-a-box modules illustrated in Figure 2.5 have been designed to be shipped by truck from the factory to the IFE plant site and to be hot-swapped while the plant continues to operate.

The modular architecture approach is essential to achieving a high operational availability for the DPSSL IFE plant. It would allow upgrades and improvements to the laser driver modules without the need for shutting down plant operation. The modular architecture would enable an IFE plant to follow an upgrade path starting with a lower plant power output and increasing plant output over time by adding banks of laser modules.

---

[44] R. Deri et al., op. cit.

[45] H. Azechi, Osaka University, "Inertial Fusion Energy: Activities and Plans in Japan," Presentation to the committee on June 15, 2011.

[46] R. Deri et al., op. cit.

[47] A. Bayramian, S. Aceves, T. Anklam, et al., 2011, Compact, efficient laser systems required for laser inertial fusion energy, *Fusion Science and Technology* 60: 28-48.

TABLE 2.1 Laser System Requirements for a Diode Laser-Pumped Solid-State IFE Driver Operating in the UV at 351 nm

Characteristic	Requirement
Total laser energy at 351 nm (MJ)	2.2
Total peak power (TW)	633
No. of beam lines	384 (48 × 8)
Energy per beam line at 351 nm (kJ)	5.4
Wall plug efficiency at 351 nm (%)	15
Repetition rate (Hz)	16
Lifetime of system (shots)	30×10^9
Availability	0.99
Maintenance (h)	<8
Beam pointing (μm root-mean-square)	100
Beam group energy stability (8 beams) (% root-mean-square)	<4
Beam-to-beam timing at target (ps root-mean-square)	<30
Focal spot (w/CCP[a]), 95% enclose (mm)	3.1
Spectral bandwidth, 3ω (GHz)[b]	180
Prepulse at 20 ns prior to main pulse (W/cm^2)	<10^8

[a] CPP, Continuous Phase Plate, is used to modify the far field from a peak to a flat top for target drive.
[b] Used for suppression of stimulated Raman scattering, stimulated Brillouin scattering, and in conjunction with a diffraction grating for smoothing by spectral dispersion (SSD) of the laser speckle induced by the use of the CPP on target.
SOURCE: A. Bayramian, S. Aceves, T. Anklam, et al., 2011, Compact, efficient laser systems required for laser inertial fusion energy, *Fusion Science and Technology* 60: 28-48.

The Global R&D Effort on Solid-State Lasers for IFE Drivers

The laser driver for IFE is a significant component (~25 percent) of the capital cost of an IFE plant and is therefore the subject of research and development aimed at maximizing the performance, availability, and reliability of DPSSL driver for IFE in Europe,[48] Japan,[49] China,[50] and the United States.

In France, the construction of the Laser MégaJoule (LMJ) project, a NIF-like, flashlamp-pumped Nd:glass laser system with a goal of 2 MJ drive energy,[51] is

---

[48] J. Collier, Ruther Appleton Laboratory, "Recent Activities and Plans in the EU and UK on Inertial Fusion Energy," Presentation to the committee on June 15, 2011.

[49] H. Azechi, Osaka University, "Inertial Fusion Energy: Activities and Plans in Japan," Presentation to the committee on June 15, 2011.

[50] J. Zhang, Institute of Physics, Chinese Academy of Sciences, and Xiantu He, Beijing Institute of Applied Physics and Computational Mathematics, "Inertial Fusion Energy: Activities and Plans in China," Presentation to the committee on June 15, 2011.

[51] J. Collier, Ruther Appleton Laboratory, "Recent Activities and Plans in the EU and UK on Inertial Fusion Energy," Presentation to the committee on June 15, 2011, and R. Garwin and D. Hammer, "Notes from Our LMJ Visit, February 26, 2011," Presentation to the committee on March 30, 2011.

nearing completion. This large, single-shot laser system is designed for physics and target studies. Recently, Russia announced its plans for ISKRA/UFL, a nearly 3 MJ fusion laser.

R&D in Europe and Japan is directed toward diode-pumped, cryo-cooled, Yb:YAG ceramic lasers. Cryocooling of Yb:YAG brings improved performance and optimum gain and power extraction.[52] Modern transparent laser ceramics were developed in Japan beginning in 1995.[53] Lasers based on ceramics were shown to perform as well as, or better than, single crystal lasers.[54] Today, ceramic laser gain media are available in the size 10 cm × 10 cm. Laser ceramics are still undergoing extensive research to improve the quality and consistency of the material. In the future, when commercial supplies of ceramic laser gain materials are available, ceramics may replace glass as the preferred laser host material in high-average-power IFE laser drivers. When laser ceramics do become available, the modular architecture of the proposed laser IFE driver may be able to accommodate the new gain media without making major changes to the IFE system.

In China, the development of IFE laser drivers is based on lamp-pumped Nd:glass lasers. The next step is to bring online by 2012-2013 the Shenguang (Divine Light) SG-III laser, which will operate frequency-tripled (like the NIF) at 351 nm for inertial confinement fusion experiments with 48 beams at 3 ns and 200 kJ total energy. The longer-range plan is to construct and operate the NIF-scale SG-IV laser by 2020 at 3 ns and 1.5 MJ (351 nm). Work has also been initiated in China on diode-pumped, cryocooled, solid-state lasers for future IFE drivers.

Scientific and Engineering Challenges and Future R&D Priorities for DPSSLs for IFE Applications

The following proposed DPSSL R&D program, as described in presentations to the committee, illustrates the key technical challenges that should be addressed to mitigate risks going forward:

- *Pulsed diode laser drivers and diode laser arrays with polarization combining.* Research on the optimized design of pulse diode laser bars and arrays of bars should be pursued to optimize diode bar efficiency and power per bar and facilitate lower production costs.

---

[52] T.Y. Fan, 2007, Cryogenic Yb^{3+}-doped solid state lasers, *IEEE Journal of Quantum Electronics* 13: 448.

[53] A. Ikesue, Y.L. Aung, T. Taira, T. Kamimura, K. Yoshida, and G.L. Messing, 2006, Progress in ceramic lasers, *Annual Review of Materials Research* 36: 397-429.

[54] K. Ueda, J.F. Bisson, H. Yagi, K. Takaichi, A. Shirakawa, T. Yanagitani, and A.A. Kaminskii, 2005, Scalable ceramic lasers, *Laser Physics* 15 : 927-938.

- *Birefringence compensation by polarization rotation and balanced gain module pumping.* The idea of birefringence compensation by use of polarization rotation and balanced thermal loading of two gain elements is well known. Polarization rotation should be experimentally tested to determine whether specifications can be met at 15 Hz and ~130 kW average power in the IR from a laser in a box.
- *The KD*P switch for optical isolation and four pass oscillator/amplifier control.*[55] The KD*P polarization switch is placed in the low optical fluence zone of the laser system. However, the KD*P must be cooled and the appropriate 20 kV electric field applied for switching. The operation of this switch should be tested to validate modeling and assure proper operation under repetition rate and thermal loading.
- *Efficiency and thermal cooling of the KD*P harmonic generation converter.* The KD*P nonlinear frequency converter operates at average power and is cooled with flowing helium gas. The conversion efficiency of the convertor and the operation at average power should be determined by testing at full average power.
- *UV beam line damage testing and beam delivery utilizing the fused silica Fresnel lens at 580°C.* The UV beam line is a critical element in the delivery of the laser power to the chamber and through the Fresnel lens to a focus at the target position. Optical damage testing should be done to assure reliable operation of the final fused silica Fresnel lens optic at operating temperature and optical fluence.
- *The laser beam-line-in-a-box should be modeled and tested at full scale.* The laser in a box is a critical element and should be tested at full scale and at operating conditions to determine if it can meet design reliability, power, pointing, and vibration and alignment requirements. It should be tested to determine that it can meet the hot-swap requirements for a line-replaceable unit.

Path Forward for Diode-Pumped Solid-State Laser-Based Inertial Fusion Energy

In this section, the integrated systems engineering and supporting R&D required to develop a solid-state, laser-driven IFE power plant is described. This plan for DPSSL drivers is based on the LIFE team's submissions to the committee and other publications.

LIFE is based on indirect-drive targets injected into a xenon-gas-filled chamber, as described in the LIFE design study. The advantages of the gas-filled chamber were

---

[55] KD*P is potassium dideuterium phosphate, a material used widely in frequency conversion optics.

described to the committee by Wayne Meier.[56] This reactor would be made of steel with a 6-m-diameter chamber comprising segmented and replaceable chamber walls. The chamber is located within the vacuum walls and is designed to be replaced periodically. The use of xenon gas reduces peak temperature spikes at the chamber walls. The 384 laser beams are focused into the indirect-drive target hohlraum through thin, heated SiO_2 Fresnel lenses protected from ion bombardment by the xenon gas. The final optics are thin to allow them to slide in and out easily during replacement and are heated to 580°C to provide self-annealing in the radiation environment. The laser propagation through the xenon gas is calculated to be acceptable at the 351 nm drive wavelength.

The R&D program must support the integrated systems engineering approach that is essential for designing a power plant facility that meets customer needs at a cost that is competitive with other sources of energy such a modern fission reactors.[57] Issues for which R&D is critical include target physics, design and cost, and survival of the target during injection and engagement at more than 1 million targets per day. Also of interest are recycling of the lead used for the hohlraum, as well as tritium breeding and control—all in addition to the development of reliable, efficient laser drivers.

Near-Term R&D Objectives (≤5 Years)

The proposed Nd-doped APG glass DPSSL driver is based on performance metrics provided by NIF, the Mercury laser system, and commercial laser performance specifications. Prudent engineering practice requires a risk-reduction program to confirm the anticipated performance of the proposed IFE laser driver design. A high-priority, near-term R&D objective is to design, build and test a full-scale laser beam-line module.[58] This single laser beam line should achieve all design specifications, including the specifications necessary for a laser line-replaceable-unit that enables a hot-swap exchange in an IFE plant environment.

The laser beam-line module demonstration would allow full-aperture and average-power testing of pulsed laser diode drivers and laser diode arrays with polarization combining. Research is needed to facilitate optimization of pulsed diode bars and arrays of bars to optimize diode bar efficiency and power per bar and to facilitate lower production costs.

---

[56] W. Meier, LLNL, "Overview of Chamber and Power Plant Designs for IFE," Presentation to the committee on January 29, 2011.
[57] T.M. Anklam, M. Dunne, W.R. Meier, S. Powers, and A.J. Simon, 2011, LIFE: The case for early commercialization of fusion energy, *Fusion Science and Technology* 60: 66-71.
[58] A. Bayramian, S. Aceves, T. Anklam, et al., 2011, Compact, efficient laser systems required for laser inertial fusion energy, *Fusion Science and Technology* 60: 28-48.

The UV beam line is a critical element for delivery of the laser power to the chamber and to the target through the fused-silica, Fresnel-lens, final optic. The final optics beam line and optical components should be tested to the limits available to confirm expected lifetimes and performance.

Conclusion 2-2: If the diode-pumped, solid-state laser technical approach is selected for the roadmap development path, the demonstration of a diode-pumped, solid-state laser beam-line module and line-replaceable-unit at full scale is a critical step toward laser driver development for IFE.

Conclusion 2-3: Laser beam delivery to the target via a UV beam line, the final optics components, and target tracking and engagement are critical technologies for laser-driven inertial fusion energy.

Medium-Term R&D Objectives (5-15 Years)

Assuming that ignition has been achieved and the full-scale laser beam line has been designed, constructed, tested, and met design criteria, work would begin on implementing the integrated system engineering design for a laser-driven Fusion Test Facility (FTF), a facility to demonstrate repetitive DT target shots and reactor-scale gain, using reactor-scale driver energy. The medium-term R&D objective is to design, build, and operate such a facility.

One proposal from the LIFE team is a solid-state laser-driven FTF that would operate at the 400 MW_e scale in bursts of increasing duration. Its goal would be to demonstrate a target gain of 60-70 and plant gain of ~5, consistent with a laser wall-plug efficiency of 15 percent in the UV. This facility size is a trade between capital cost and operational capability that would inform the IFE community about key aspects of plant operation and material issues in the relevant environment. It would require a chamber capable of operating for the required number of tests and a target factory capable of producing and delivering targets at the necessary rate. The most highly leveraged elements of this facility are the target chamber structural material, the target cost, and target gain,[59] so that optimization of these elements would be the key objective. The laser driver and its critical components—laser diodes, design for high efficiency, and the APG glass gain medium—are not high on the list of items that lead to a large variance in the cost of electricity.[60]

---

[59] T.M. Anklam, M. Dunne, W.R. Meier, S. Powers, and A.J. Simon, 2011, LIFE: The case for early commercialization of fusion energy, *Fusion Science and Technology* 60: 66-71.
[60] Ibid.

The FTF would be designed such that it could be upgraded to the 1 GW$_e$ power output level in the future. The key issues in moving forward are a combination of technical issues and licensing issues associated with the plant operation and integrated facility design.[61]

The technologies that would be demonstrated at the FTF include:

- Laser system;[62]
- Integrated facility design;[63]
- Target production, injection, and engagement;[64]
- Chamber and blanket design;[65]
- Thermoelectric plant; and
- Tritium plant.

Success of a laser-driven facility and the projection of the technology to a cost-effective power plant would assure that this technical approach is a candidate for upgrade to the demonstration-scale power plant described in Chapter 4.

Conclusion 2-4: Laser-driven inertial fusion for energy production requires an integrated system engineering approach to optimize the cost and performance of a Fusion Test Facility followed by a demonstration plant.

Long-Term R&D Objectives (>15 Years)

The long-term objectives are to define a path for commercial energy production based on IFE. The goal can be met if the 400 MW$_e$ FTF leads to a 1 GW$_e$ power plant facility 10 to 15 years following completion of the FTF.

The details of the progression in the design and performance for each stage of the roadmap to the demonstration facility and then to the commercial power plant have been described by Tom Anklam. Table 2-2 (taken from Anklam's presentation) shows a conceptual roadmap for a commercialization path that has

---

[61] W. Meier, LLNL, "Overview of Chamber and Power Plant Designs for IFE," Presentation to the committee on January 29, 2011.

[62] A. Bayramian, S. Aceves, T. Anklam, et al., 2011, Compact, efficient laser systems required for laser inertial fusion energy, *Fusion Science and Technology* 60: 28-48.

[63] M. Dunne, E.I. Moses, P. Amendt, et al., 2011, Timely delivery of laser inertial fusion energy (LIFE), *Fusion Science and Technology* 60: 19-27.

[64] R. Miles, M. Spaeth, K. Manes, et al., 2011, Challenges surrounding the injection and arrival of targets at LIFE fusion chamber center, *Fusion Science and Technology* 60: 61-65.

[65] J.F. Latkowski, R.P. Abbott, S. Aceves, et al., 2011, Chamber design for the laser inertial fusion energy (LIFE) engine, *Fusion Science and Technology* 60: 54-59.

TABLE 2.2 Conceptual Roadmap for the Commercialization Path for LIFE

Design/Performance	LIFE 1	LIFE 2	LIFE 3
Laser energy 3ω (MJ)	1.3	2.4	2.0
Repetition rate (Hz)	14.8	14.8	14.8
Plant electrical gain	1.3	4.4	7.0
House power fraction[a]	0.77	0.25	0.16
Thermal-to-electric efficiency (%)	43	48	53
First wall material[b]	RAFMS	ODS	ODS
Radius (m)	3.7	5.6	6.2
First wall neutron loading lifetime (full power equivalent)			
(MW/m^2)	1.9	4.5	4.5
(dpa/yr)	20	50	50
(yr)	0.9	4.5	4.5
Fusion yield (MJ)	27	147	180
Target gain	21	64	94
Fusion power (MW)	400	2,200	2,660
Availability allocation[c] (%)	50	92	92

[a] Also known as recirculating power fraction.
[b] RAFMS is a low-activation ferritic/martensitic steel and ODS is an oxide dispersion strengthened steel.
[c] The availability allocation is not a bottom-up calculation but is used to set targets for the LIFE subsystems in regard to reliability, replacement time, and redundancy.
SOURCE: T.M. Anklam, LLNL, "LIFE Economics and Delivery Pathway," Presentation to the committee on January 29, 2011.

been proposed.[66] It consists of three stages. The first stage, referred to as LIFE 1, is the 400 MW$_e$ facility described above and is based on the 384 laser module design. LIFE 1 is projected to be operational 10 to 15 years following ignition on NIF at a total build cost of between $4 billion and $6 billion. LIFE 1 will provide operational capability similar to a commercial power plant and will provide the fusion environment required for testing materials in the relevant environment. LIFE 1 is designed to allow an upgrade in scale to the 1 GW$_e$ demonstration power plant referred to as LIFE 2 in Table 2-2. The learning curve would lead to an improvement in plant performance at a cost similar to the first plant. The third step, referred to as LIFE 3 power plant design, captures the improvements gained from LIFE 2 operation and provides insight into the economics for the commercial power plant operation.

---

[66] T.M. Anklam et al., op. cit.

Krypton Fluoride Lasers

Background and Status

The krypton fluoride laser is an excimer laser that radiates in a broad, 3-THz band at the deep ultraviolet wavelength of 248 nm. In high-energy applications, its gaseous laser medium containing argon, krypton, and less than 1 percent fluorine is pumped by electron beams. Because inductance slows the rise of high-current electron beams and the excimer upper-state radiative lifetime is only on the order of 1 ns in typical conditions, the "angular multiplex" architecture was proposed[67] to compress electron beam energy delivered in several hundred nanoseconds down to a laser fusion driver pulse of few nanoseconds. The multiplex architecture passes many sequential copies of the desired drive pulse through the electron-beam-pumped medium, extracting all of the energy, before the copies are time-shifted to all arrive simultaneously at the target.

In the mid-1980s, seminal work was reported on the increased stability[68] and drive efficiency[69] of direct-drive laser fusion with the use of deep UV laser light (at 250 nm) as opposed to the 1 μm (or longer) wavelength used previously. As the various laser-plasma instabilities were studied in more detail, their intensity thresholds were mainly found to increase with decreasing wavelength, motivating the transition of laser fusion experiments to the third harmonic of the neodymium glass laser (351 nm) or the krypton fluoride (KrF) laser (248 nm). With higher instability thresholds, the achievable acceleration of the target was increased. The technique of incoherent spatial imaging (ISI)[70] was introduced to provide uniform and broad-band illumination and to further suppress acceleration instabilities. The electron-beam-pumped KrF gas laser was an excellent fit to requirements, with a wavelength of 248 nm and a 3 THz bandwidth to suppress laser–plasma instabilities. The first moderate-energy (5 kJ) KrF laser design—called Nike—was built at the Naval Research Laboratory (NRL) in the early 1990s. This was a single-shot facility without gas recirculation. Under the High Average Power Laser (HAPL) program (see Chapter 1), a 5 Hz, 700 J KrF laser called Electra was built and tested (see Figure 2.6). With Electra, the KrF laser technology was demonstrated

---

[67] J.J. Ewing, R.A. Haas, J.C. Swingle, E.V. George, and W.F. Krupke, 1979, Optical pulse compressor systems for laser fusion, *IEEE Journal of Quantum Electronics* QE-15: 368-379.

[68] M.H. Emery, J.H. Gardner, and S.E. Bodner, 1986, Strongly inhibited Rayleigh-Taylor growth with 1/4 micron lasers, *Physical Review Letters* 57: 703-706.

[69] J.H. Gardner and S.E. Bodner, 1986, High-efficiency targets for high-gain inertial confinement fusion, *Physics of Fluids* 29: 2672-2678.

[70] R.H. Lehmberg and S.P. Obenschain, 1983, Use of induced spatial incoherence for uniform illumination of laser fusion targets, *Optics Communications* 46: 27-31.

FIGURE 2.6 The 5 Hz, 700 J Electra laser at the Naval Research Laboratory. SOURCE: J.D. Sethian and S.P. Obenschain, "Krypton Fluoride Laser Driven Inertial Fusion Energy," Presentation to the committee on January 29, 2011. See also J.D. Sethian et al., 2010, The science and technologies for fusion energy with lasers and direct drive targets, *IEEE Transactions on Plasma Science* 3: 690-703.

and supported with modeling at a scale to support KrF as a technical application approach for an IFE laser driver.

The KrF laser is suitable to illuminate direct-drive targets because of its UV wavelength. However, the projected 7 percent efficiency of the KrF laser requires a target gain of more than 140. For conventional direct-drive targets this would require a laser drive energy of 2.4 MJ. One strategy to decrease the drive energy is to use high-velocity direct drive.[71] In this case, the required drive energy is calculated to be near 1 MJ. A second strategy, which would be more attractive if it is feasible, is to use relatively low driver energy to provide compression and to achieve ignition by applying a late but very high-peak-power shock ignition pulse (see Figure 2-2). Shock ignition, similar to fast ignition (see Figure 2-1), is attractive for laser-based inertial fusion energy because it may potentially decrease the driver energy by a factor of 5 from ~2 MJ (conventional direct drive) to approximately 0.4 MJ. However, it should be noted that neither fast ignition nor shock ignition has been explored

---

[71] S. Obenschain et al., 2006, Pathway to a lower cost high repetition rate ignition facility, *Physics of Plasmas* 13: 056320.

experimentally at the drive energies relevant for ignition. How driver size affects the capital cost of a plant and the cost of electricity is discussed in Chapter 3.

The homogeneous bandwidth of KrF is 3 THz; consequently, strongly time-randomized beams[72] may be used to suppress laser–plasma instabilities. Theory predicts potential suppression of a particular instability when the laser coherence length becomes shorter than the relevant plasma scale length, which itself increases the thresholds: for example, for stimulated Brillouin scattering (SBS), the plasma velocity gradient, and for stimulated Raman scattering (SRS), the plasma density scale length.

The optical system of a KrF laser fusion amplifier focuses an incoherent KrF light source at the laser "front end" onto the target. This technique, called incoherent spatial imaging, allows a uniform intensity profile on the target, essential for acceleration with minimum growth of instabilities. Uniform irradiation has been demonstrated with KrF laser beams at NRL.[73] Simulations of high-gain, direct-drive targets[74] include the appropriate KrF spectrum of intensity fluctuations, modified to account for the typical number (approximately six) of overlapping beams at any point on the target surface.

The same optical design also allows dynamic focusing on a compressing target—or "zooming"—to improve efficiency by matching the focal spot to the shrinking pellet size during compression. This works by switching successively smaller incoherent source images into the front end of the laser. As the front end is imaged onto the target, the decrease in target size can be matched. Zooming has been demonstrated on the NRL Nike laser. It is calculated that approximately 1.5 times less laser energy is required to achieve fuel compression when zooming is employed.[75]

The KrF angular multiplexing geometry is well-suited for the generation of sub-nanosecond shock pulses, which can be done without any efficiency penalty, according to complete laser kinetic modeling.[76] This works because the 0.2-ns shock spike extracts energy that has been stored in the KrF medium on the 1 ns timescale. Separate angular multiplex paths ensure that the full spike intensity is not experienced on any optical surface prior to synchronous arrival at the target, decreasing

---

[72] Intensity smoothing on a short timescale via the high frequency of fluctuations inherent in beams of high bandwidth.

[73] J.D. Sethian and S.P. Obenschain, "Krypton Fluoride Laser Driven Inertial Fusion Energy," Presentation to the committee on January 29, 2011.

[74] A.J. Schmitt, 1984, Absolutely uniform illumination of laser fusion pellets, *Applied Physics Letters* 44: 399-401.

[75] S.P. Obenschain and A.J. Schmitt, NRL, Presentations to the Target Physics Panel on September 20, 2011.

[76] R.H. Lehmberg, J.L. Giuliani, and A.J. Schmitt, 2009, Pulse shaping and energy storage capabilities of angularly-multiplexed KrF laser fusion drivers, *Journal of Applied Physics* 106: 023103.

substantially the risk of optical damage. Because the 248 nm light is generated from the outset in the KrF medium, there is no need to frequency convert at the final optical stage via intensity-dependent nonlinear optical crystals that have limited dynamic range.

A beneficial feature for repetition rate operation of a gas medium in a KrF laser is that the waste heat is carried away by circulating the gas. Further, the gaseous laser medium is self-healing in the face of optical damage. The multiplexed beams propagate at approximately 100 times the diffraction limit and so are not significantly distorted by residual refractive index variations in the gas.

The wall-plug efficiency of a KrF laser is expected to exceed 7 percent, based on individual components that have been demonstrated at NRL. The separate demonstrations involve durable, solid-state pulsed power; guided electron-beam transmission through the foil support structure; and optical extraction. Although all components have not yet been demonstrated in a single device, these are separable efficiencies that multiply to generate the anticipated 7 percent efficiency. After nearly 10 years of development, KrF has delivered runs of 5×10^4 pulses at 5 Hz (~3 h) and 1.5×10^5 pulses at 2.5 Hz (~17 h) with 270 J/pulse.[77]

Scaling of KrF laser energy from its present 5 kJ to the 20 kJ module needed for a power plant has been the subject of detailed theoretical study.[78] Designs up to more than 50 kJ appear possible. In a 400 kJ facility, for example, 20 of the basic 20 kJ modules would be required. Continuous plant operation could be possible via the type of architecture proposed for the KrF FTF,[79] in which spare modules can be switched into use by rotating the mirror a few degrees at the entry and exit of common beam transport ducts. The electron beams that drive the KrF gain medium can also be designed modularly for ease of substitution.

Scientific and Engineering Challenges and Future R&D Priorities for KrF Lasers for IFE Applications

The following are key KrF laser R&D priorities for the future, as described in presentations to the committee:

- *Laser–plasma instabilities.* This is discussed earlier in the chapter.

---

[77] J. Sethian and S. Obenschain, Naval Research Laboratory, "Krypton Fluoride Laser Driven Inertial Fusion," Presentation to the committee on January 29, 2011.

[78] R.H. Lehmberg, J.L. Giuliani, and A.J. Schmitt, 2009, Pulse shaping and energy storage capabilities of angularly-multiplexed KrF laser fusion drivers, *Journal of Applied Physics* 106: 023103, and references therein.

[79] S.P. Obenschain, J.D. Sethian, and A.J. Schmitt, 2009, A laser based fusion test facility, *Fusion Science and Technology* 56: 594-603.

- *The KrF laser lifetime, energy scale, pulse shaping, and optics.* During the development of the Electra 5 Hz KrF laser at NRL, the solutions to integrated engineering challenges were demonstrated by system runs of greater than 10^5 pulses.[80] Demonstrations still need to be extended to beyond 1.6×10^8 pulses (one year at 5 Hz). The electron gun cathode is a critical element that has been demonstrated to greater than 5×10^5 pulses (to date), and a prototypical solid-state, pulsed-power module has been tested to more than 10^7 pulses. The fatigue life of the foil barrier between the electron gun and the laser gas is theoretically sufficient for more than 10^8 pulses (at 370°C). Fatigue has not been a principal concern, but the foil life has been limited by reverse arcs that occur postpulse within the electron gun.[81] Elimination of these arcs by tuning has extended the foil life to more than 10^5 pulses.[82] Gas switches in the pulsed-power supply currently limit runs to 10^5 pulses, because they generate voltage spikes that cause that arcing. This problem is removed with solid-state pulsed power, which has already been demonstrated separately to more than 10^7 pulses, as noted above. The overall laser engineering challenge is to extend demonstrations from the greater than 10^5 level to the greater than 1-year level, and to understand the statistics of failure.
- *The energy of a single module of the KrF laser.* This is projected to scale to at least 16 kJ from existing systems.[83] Higher module energy, up to 30 kJ, may be possible.[84] In regard to the "front end" of the laser, where pulse shaping is done, NRL has identified[85] a nonlinear optical process to transfer fiber laser waveforms (already well developed for the NIF laser system) to drive the KrF laser system. The bandwidth of the fiber laser system is 0.5 THz and the timing accuracy is 30 ps. It has been shown by detailed calculation that arbitrary shock ignition waveforms may be generated without an efficiency penalty in a KrF amplifier,[86] although this has to be confirmed experimentally. Demonstration of "end-to-end" wall plug efficiency of 7 percent is an important development objective.

---

[80] J. Sethian and S. Obenschain, Naval Research Laboratory, "Krypton Fluoride Laser Driven Inertial Fusion," Presentation to the committee on January 29, 2011.

[81] Ibid.

[82] Ibid.

[83] Ibid.

[84] R.H. Lehmberg, J.L. Giuliani, and A.J. Schmitt, 2009, Pulse shaping and energy storage capabilities of angularly-multiplexed KrF laser fusion drivers, *Journal of Applied Physics* 106: 023103, and references therein.

[85] Ibid.

[86] Ibid.

- *The degradation of the laser windows by laser gas and the lifetime of the final optics are two challenges for the KrF driver optics.* The first challenge deals with the slow degradation of the fused silica laser windows by the laser gas or, possibly, by moisture contamination within it. There are fallback approaches in which a fluorine-depleted gas layer is deployed next to the window or silica windows are changed to calcium fluoride. However, attention to gas purity and dryness may also solve the problem. The committee notes the commercial achievement of billion-pulse lifetimes in sealed KrF lasers for lithography. As for the second challenge, the final grazing-incidence metal mirror has not yet been fabricated or exposed to fusion neutrons. It will have to be composed of materials that are stable to moderate neutron flux. Designs have been developed that minimize its neutron exposure,[87] and dielectric mirrors[88] that are radiation-resistant have exhibited good optical damage resistance at 248 nm, even after irradiation. Further irradiation and damage testing is needed on optical elements that could serve as a plasma-facing final optic. Dielectric mirrors may qualify for this function. A magnetic field is probably required to divert fast ions before they can impact a final mirror, although X-ray energy bursts must also be withstood. Designs for magnetic field "intervention" have been proposed.[89]

Conclusion 2-5: The demonstration of a reactor-scale KrF module with a pulse count (before servicing) three orders of magnitude greater than presently achieved remains challenging. A key to achieving this goal would be integrating a solid state switching system into the Electra KrF laser at NRL.

Conclusion 2-6: If the KrF laser technical approach is selected for the roadmap development path, a very important element of the KrF laser inertial fusion energy research and development program would be the demonstration of a multi-kilojoule, 5-10 Hz, KrF laser module that meets all of the requirements for a Fusion Test Facility.

The timing for this step is discussed in Chapter 4.

A key R&D priority for the future is to conduct spherical direct-drive experiments using ganged 20 kJ KrF modules. The acceleration stability of 248-nm-irradiated

---

[87] L.L. Snead, K.J. Leonard, G.E. Jellison Jr., M. Sawan, and T. Lehecka, 2009, Irradiation effects on dielectric mirrors for fusion power reactor application, *Fusion Science and Technology* 56: 1069-1077.
[88] Ibid.
[89] J.D. Sethian, NRL, "The Science and Technologies for Fusion Energy with Lasers and Direct-Drive Targets," Presentation to the committee on June 15, 2011; J.D. Sethian et al., 2010, The science and technologies for fusion energy with lasers and direct drive targets, *IEEE Transactions on Plasma Science* 38(4): 690-703.

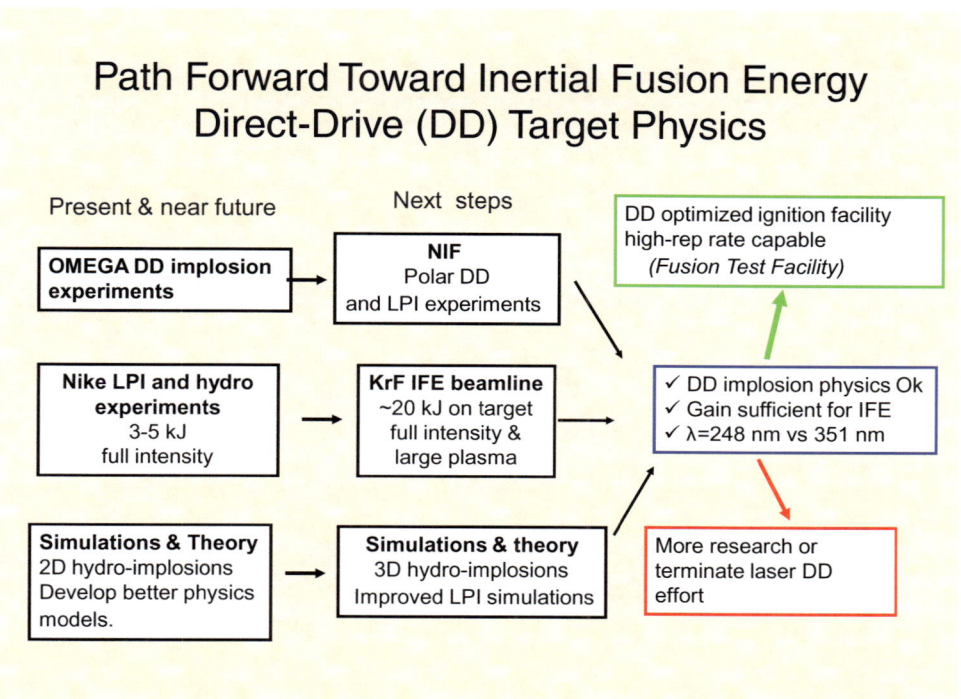

FIGURE 2.7 Diagrammatic laser IFE roadmap for direct-drive target physics research to prepare for an FTF. SOURCE: J.D. Sethian and S.P. Obenschain, NRL, "Krypton Fluoride Laser Driven Inertial Fusion Energy," Presentation to the committee on January 29, 2011.

targets may be studied initially with 1-steradian segments of target and a single 20 kJ module as proposed below by the NRL (Figure 2.7), giving information at the precise intensity and scale lengths relevant to 240 kJ implosions. The effect of target design changes for different adiabats could then be understood in detail. With good results at this energy level, four or eight 20 kJ modules could be combined in order to refine the comparison of experiment to theory, particularly in regard to the shock ignition regime at 10^{16} W/cm^2. Supporting the use of a relatively small number of beams is the Schmitt theorem on perfectly uniform illumination.[90] With zooming, the Schmitt cosine-squared intensity profile can be adjusted to the decreasing pellet size during compression, maintaining uniformity.

---

[90] A.J. Schmitt, 1984, Absolutely uniform illumination of laser fusion pellets, *Applied Physics Letters* 44: 399-401.

Path Forward for KrF Laser-Based IFE

Figure 2-7 outlines a path forward for exploration of laser direct-drive target physics involving both solid-state and KrF laser drivers. The plan for KrF laser drivers that immediately follows it is based on the NRL submission to the committee, with the exception, as noted, of ganged 20 kJ modules for exploration closer to reactor scale.

Near-Term R&D Objectives (≤5 Years)

Subscale components would be as follows:

- Convert Electra repetitive KrF facility to solid-state pulsed power (path known).
- Develop front-end discharge amplifier (design available) and build pulse-shaper.
- Design and test components for prototype 20 kJ module, initially at 0.01 Hz.
- Refine target design and physics.
- Complete efforts on other IFE technologies begun in the HAPL program:

  —Chamber physics (engineered walls, magnetic intervention);
  —Chamber technology (blanket, neutronics);
  —Materials (experimental and theoretical);
  —Final optics (grazing incidence metallic mirrors, dielectrics);
  —Target fabrication (shells, layering); and
  —Target injection and tracking.

The cost guidance for this Phase I (estimate provided by NRL) was as follows. For the KrF target physics and laser development alone, approximately $25 million per year would be required over 3-4 years. A program that included development of essential auxiliary technologies (target fabrication, fusion materials, and system studies to provide guidance) would cost about two to three times that amount. As a point of comparison, the HAPL program peaked at $25 million per year in 2006.

Medium-Term (5-15 Years)

Phase II would consist of a full-size KrF laser beam line (20 kJ at 5 Hz) along with other inertial fusion energy components. As shown in Figure 2.7, the following steps assume testing of polar direct drive on the NIF:

- Build and test 20 kJ, 5 Hz beam line.

- Engage targets injected into test chamber with beam line.
- Develop all critical inertial fusion energy technologies (e.g. low-cost targets, full-size final optics) for the Fusion Test Facility.
- Develop high confidence in pellet designs and physics (using NIF and KrF beam line).

The cost guidance for this Phase II (provided by NRL) estimates that $50 million per year over 5 years would enable development of a full-scale KrF beam line for the FTF and demonstration of highly reliable operation. The overall Phase II program would require about $150 million to $200 million per year to develop all the required technologies for the FTF and to design it. Additional ganged 20 kJ modules for higher energy target experiments will cost between $10 million and $20 million each, over and above the NRL estimated Phase II cost.

Long-Term R&D Objectives (>15 Years)

The FTF with a 500 kJ KrF laser that will do the following:

- Show that IFE components routinely perform with precision and durability.
- Optimize target performance.
- Develop, test, and qualify fusion materials and components.
- Demonstrate reliable operation with nominal 250 MW fusion power.
- Attract significant participation by private industry.
- Provide the technical and cost basis for full-scale power plants.

It is too early to develop reliable cost estimates for the Phase III work of building and operating the FTF. Use of a KrF driver is predicted to reduce the driver energy requirements substantially, with a beneficial impact on the cost.

Heavy-Ion Accelerators

Background and Status

The U.S. Department of Energy supported the development of heavy-ion accelerators for fusion power production until 2003, and it funded several conceptual power plant designs for both accelerator and laser drivers. The most recent conceptual design for a heavy-ion power plant[91] used an induction linear accelerator (linac), ballistic neutralized focusing, a thick liquid-protected wall, and an

---

[91] S.S. Yu, W.R. Meier, R.P. Abbott, et al., 2003, An updated point design for heavy ion fusion, *Fusion Science and Technology* 44: 266-273.

indirectly driven target. This design utilized singly charged bismuth ion beams at ≤4 GeV, an accelerating gradient ≤1.5 MV/m, and a linac length exceeding 3 km. The total beam energy was 7 MJ with target gain of 60. The linac was based on standard components: warm-bore, superconducting quadrupole magnets, thyratron pulsers, and then available ferromagnetic materials for the induction cores.

The most recent two-dimensional simulations of indirectly driven targets, carried out by LLNL, showed better performance than the target used for the conceptual power plant design. Specifically, the simulations indicated that it would be possible to achieve gains on the order of 90 to 130 at beam energies from 1.8 to 3.3 MJ, respectively.[92] The two-dimensional codes used were the same as those used for laser drivers, but the X-rays were produced when the ion beams hit material inside the hohlraum rather than the hohlraum walls, as with laser beams. NIF tests should lead to a better understanding of the performance of such indirect targets.[93]

There are multiple accelerator options for heavy-ion fusion (HIF). The two most promising options are induction accelerators and radio-frequency (RF) accelerators. There has not been sufficient funding to develop both options in the United States. For more than two decades, there has been an informal understanding that Europe and Japan would pursue the RF option while the United States would pursue the induction option. The largest foreign programs are based on existing or planned multipurpose RF accelerators using storage rings. Since these accelerators are multipurpose machines, they are not ideally matched to some of the requirements of IFE. Nevertheless, the largest of the new machines, the Terawatt Accelerator (TWAC) at the Institute for Theoretical and Experimental Physics in Moscow and FAIR at GSI in Darmstadt, will be much more able to create high temperatures and high pressures (pressure predicted to be in the 1-100 Mbar regime) than existing U.S. induction accelerators.[94] TWAC is currently under construction, and ground has just been broken for FAIR.

In addition to the foreign programs, the privately funded Fusion Power Corporation in the United States has been exploring the possibility of using radio-frequency technology without storage rings to power multiple reaction chambers.[95]

Beneficial Features of Heavy-Ion Fusion

Heavy-ion drivers have a number of beneficial characteristics:

---

[92] D. Callahan-Miller and M. Tabak, 2000, Progress in target physics and design for heavy ion fusion, *Physics of Plasmas* 7: 2083-2091.

[93] J.D. Lindl, P. Amendt, R.L. Berger, et al., 2004, The physics basis for ignition using indirect-drive targets on the National Ignition Facility, *Physics of Plasmas* 11: 339-491.

[94] B. Sharkov, FAIR, "Heavy Ion Inertial Fuse Ion Energy—Activities in Europe and in Russia," Presentation to the committee on October 31, 2011.

[95] C. Helsley, Fusion Power Corporation, Presentation to the committee on February 22, 2012.

- High-energy particle accelerators of megajoule-scale beam energy have separately exhibited efficiencies, pulse rates, average power levels, and durability required for IFE.
- The relatively high efficiency permits the use of indirect drive, and liquid walls can be used because the high-energy beams can penetrate through high vapor pressure caused by the hot liquid.
- Heavy ions deposit their energy within the case volume. The cases protect the fuel capsules as they move toward the center of a hot reaction chamber.

Recent Successes

In recent years, the program has been undertaken by a Virtual National Laboratory consisting of LBNL, LLNL, and the Princeton Plasma Physics Laboratory (PPPL), with additional work at the University of Maryland.

- The Single Beam Transport Experiment demonstrated that space-charge-dominated beams could be transported without emittance growth, as required for heavy-ion fusion. Emittance growth degrades the ability to focus the beam. If emittance growth were excessive, heavy ion fusion would not be feasible.
- Multiple-beam experiments addressed acceleration, current amplification, longitudinal confinement, and multibeam transport. The high-current experiment studied driverlike beam transport. The three-dimensional particle simulations using WARP code modeled secondary electrons successfully.
- Beam transport with driver-scale line-charge density and without emittance growth was demonstrated.
- Beams were compressed from 500 ns to a few nanoseconds in the neutralized drift compression experiment-1 (NDCX-I).
- Beams were focused to millimeter spot size using innovative plasma sources.
- An end-to-end numerical simulation capability was developed.

Scientific and Engineering Challenges and Future R&D Priorities for Heavy-Ion Accelerators for IFE Applications

As is the case for nearly all credible fusion options, the projected cost of electricity in earlier studies[96] was higher than the cost for many existing power options

---

[96] S.S. Yu, W.R. Meier, R.P. Abbott, et al., 2003, An updated point design for heavy ion fusion, *Fusion Science and Technology* 44: 266-273; DOE, 1992, *OSIRIS and SOMBRERO Inertial Fusion Power Plant Designs*, Final Report, DOE/ER/54100; DOE, 1992, *Inertial Fusion Energy Reactor Design Studies, PROMETHEUS-L and PROMETHIUS-H*, Final Report, DOE/ER/54101. More recent design studies that have been reviewed as rigorously as those cited here do not exist in this case.

such those based on fossil fuels or fission. However, the projected cost of electricity was usually lower with heavy-ion fusion than with the laser option, partly because of the comparatively high efficiency of heavy-ion drivers (calculated to be 25 to 40 percent).[97] It should be noted that large accelerators often exceed the repetition rate required for IFE. For example, the Spallation Neutron Source operates at 60 Hz, and intershot switching might allow operation with multiple chambers. Nevertheless, cost reduction remains an important challenge. The cost of the accelerator decreases with decreasing target energy and more relaxed requirements on beam quality and alignment tolerances. For this reason, a cost reduction program should include improved target designs. There has been significant progress in this area.[98] Also, prior to its termination in 2003, the HIF program had initiated a multipronged program to reduce the cost of accelerators. This program included development of the following:

- Inexpensive, compact, long-life ion sources.
- Compact, quadrupole magnet arrays amenable to robotic assembly or other mass production techniques. Some cold-bore quadrupole designs used a cooled liner, similar to Large Hadron Collider technology.[99] This technology was expected to lead to smaller, less expensive accelerators than the warm-bore option.
- High-gradient insulators cast from glassy ceramics or fabricated from other materials. The object was to reduce manufacturing costs and increase the acceleration gradient to reduce the length and cost of the accelerator.
- Advanced solid-state pulsers using technology similar to that proposed for KrF lasers and pulsed-power fusion.
- Better ferromagnetic materials. This effort involved working with vendors to reduce the cost of newly developed, low-loss materials and interlaminar insulation techniques.

Although the cost reduction program and other parts of the program aimed at fusion energy were discontinued in 2003, accelerator development was fortunately able to continue at a modest budget level in support of high-energy-density physics research. Most recently, Recovery Act funds have allowed the construction of the NDCX-II accelerator. Because NDCX-II incorporates some features of a power plant driver, albeit at small scale, it provides a very good test bed for the validation

---

[97] See the DOE reports listed in the preceding reference.

[98] D. Callahan-Miller and M. Tabak, 2000, Progress in target physics and design for heavy ion fusion, *Physics of Plasmas* 7: 2083-2091.

[99] O. Groebner, The LHC vacuum system, *Proceedings of the 1997 Particle Accelerator Conference*, IEEE Catalog Number 97CH36167, p. 3542.

of theory and simulation. While NDCX-II is not the ideal first step if IFE rather than high-energy-density physics research is the primary goal, it will help to resolve some of the critical issues needed to determine heavy-ion fusion's feasibility.

Two important requirements for IFE are high repetition rates and driver durability. In regard to these requirements, existing large accelerators often meet or exceed fusion requirements.[100] For example, the average beam power in large storage rings can readily exceed 1 TW.[101] Specific challenges include the following:

- Demonstrating the projected HIF accelerator efficiency of 25 to 40 percent. Note that existing accelerators have a maximum efficiency of 12 percent, but studies in Europe, India, and the United States (of radio-frequency accelerators) suggest that between 37 and 45 percent is possible.[102]
- Narrowing the uncertainty in the attainable accelerating electric field gradient.
- Developing long-life ion sources and the other reliable and durable accelerator technologies noted above. These developments are needed to provide reliable data on efficiency and cost, and for defining the acceptable level of trips and the necessary redundancy to accommodate them.
- Optimizing plasma source development technology for intense ion-beam pulse compression and focusing.
- Raising the beam energy from ~1 J to ~100 kJ per beam. The voltage must be increased from 10 MeV to a few GeV, and the beam current must be increased from amperes to ~kiloamperes per beam.

---

[100] See J. Jowett, 2011, "Heavy Ions in 2011 and Beyond, Chamonix," 2011 LHC Performance Workshop, January 2-28, available at http://indico.cern.ch/conferenceOtherViews.py?view=standard&confId=103957; R.S. Moore, "Review of Recent Tevatron Operations," *Proceedings of the Particle Accelerator Conference 2007*, available at http://accelconf.web.cern.ch/AccelConf/p07/PAPERS/TUOCKI01.PDF; L. Rivkin, (LPAP) "PSI Sets World Record with 1.4 MW Proton Beam," available at http://actu.epfl.ch/news/psi-sets-world-record-with-14-mw-proton-beam/; M. Seidel, S. Adam, A. Adelmann, et al., 2010, "Production of a 1.3 MW Proton Beam at PSI," *Proceedings of IPAC'10* p.1309-1313, available at http://accelconf.web.cern.ch/accelconf/IPAC10/papers/tuyra03.pdf; T. Hardek, M. Crofford, Y. Kang, S-W Lee, et al., "Status of the Oak Ridge Spallation Neutron Source (SNS) RF Systems," available at http://accelconf.web.cern.ch/AccelConf/PAC2011/papers/thoas3.pdf; K. Takayama and R.J. Briggs (eds.), 2011, "Induction Accelerators," *Particle Acceleration 7 and Detection*, DOI 10.1007/978-3-642-13917-8_2, Springer-Verlag, available at http://www.springer.com/physics/particle+and+nuclear+physics/book/978-3-642-13916-1.

[101] S. Myers, "Four Decades of Colliders (from the ISR to LEP to the LHC)," *Proceedings of IPAC'10*, Kyoto, Japan, available at http://accelconf.web.cern.ch/AccelConf/IPAC10/papers/thppmh03.pdf.

[102] S.S. Kappoor, 2002, Accelerator-driven sub-critical reactor system (ADS) for nuclear energy generation, *Indian Academy of Sciences* 59: 941; and B. Aune et al., 2001, "SC Proton Linac for the CONCERT Multi-Users Facility," Particle Accelerator Conference, 2001.

- Refining the designs of the final optics and focusing system for reactor-level beams.
- Developing and testing targets that have lower input energy requirements.
- Demonstrating technologies needed to produce repetitively cycled liquid walls.

The committee notes as follows:

- While the base case considered for HIF uses an induction linac, indirect drive, and thick liquid walls, other options are possible, such as polar direct drive, shock ignition, and thin liquid or solid walls. Polar direct drive is an option that is currently being studied for both lasers and ion beams. If direct drive is successful, it is expected to have lower energy requirements and higher gain than indirect drive. Moreover, polar illumination with heavy-ion beams is compatible with the thick liquid wall chambers. These chambers minimize material damage problems.
- The final optics in HIF can be shielded from the neutrons, and neutronics calculations indicate lifetimes ≥100 years.[103] However, if the option of neutralized ballistic transport with in-vessel plasma sources were to be used, additional analysis would be required in regard to the plasma sources.
- Fast ignition and other target options, such as the X-target,[104] are being studied.[105] As a matter of historical interest, the first target considered for HIF was based on fast ignition.[106]

Path Forward for Heavy-Ion Accelerator-Based IFE

The plan for HIF IFE that follows is based on information provided to the committee by LBNL.

Near Term (≤5 Years)

- Continue the program in high-energy-density physics on the NDCX-II facility.

---

[103] J.F. Latkowski and W.R. Meier, 2003, Shielding of the final focusing system in the robust point design, *Fusion Science and Technology* 44: 300.

[104] See Figure 2-6 in the Target Physics Panel report for an image of the x-target.

[105] G. Logan, LBNL, "Heavy Ion Fusion," Presentation to the committee on January 29, 2011, and personal communication with D. Lang, NRC, in June 2011.

[106] A.W. Maschke, 1975, Relativistic ions for fusion applications, Proceedings of the 1975 Particle Accelerator Conference, Washington, D.C., *IEEE Transactions on Nuclear Science* NS-22 (3): 1825.

- Show agreement with benchmark simulations and end-to-end simulation in NDCX-II.
- Continue the collaboration with foreign heavy ion accelerator programs.

Conclusion 2-7: Demonstrating that the Neutralized Drift Compression Experiment-II (NDCX-II) meets its energy, current, pulse length, and spot-size objectives will be of great technical importance, both for heavy-ion inertial fusion energy applications and for high-energy-density physics.

It is important to recognize that the high-energy-density physics program, including NDCX-II, is, by itself, not a fusion energy program. Indeed, a number of program elements needed for an IFE program would have to be added:

- Restart the High-Current Experiment (HCX) (see Figure 2.8) accelerator to complete driver-scale beam-transport experiments that were dropped when the HIF program was terminated in 2003. These would include emittance evolution, electron clearing, and dynamic vacuum control in quadrupoles at 5 Hz. The HCX was designed to be close to driver scale in important parameters such as beam size, charge density, and pulse length. Furthermore, the lattice technology closely approximates fusion driver technology. Funding required[107] is ~$1.5 million for the first year and up to $8 million in subsequent years, which includes some of the enabling technology.[108]
- Restart the enabling technology development, including magnet arrays, pulsers, and the other technologies listed earlier in this section. This will provide the information needed to address issues of efficiency, cost, maintenance, and reliability. In particular, the projected efficiency of 25 to 40 percent and gradients >1.5 MV/m require experimental validation.

Conclusion 2-8: Restarting the High-Current Experiment to undertake driver-scale beam transport experiments and restarting the enabling technology programs are crucial to reestablishing a heavy-ion fusion program.

- Carry out scaled liquid-chamber experiments. HIF and the pulsed-power approaches to fusion appear to be the most likely driver technologies to allow the use of thick liquid walls.

---

[107] As estimated by G. Logan in a presentation to the committee in January 2011.

[108] According to G. Logan (ibid.), this is an absolute minimum budget to restart the HIF program. A higher level of funding would be required to move the program expeditiously if a vigorous inertial fusion energy program is supported.

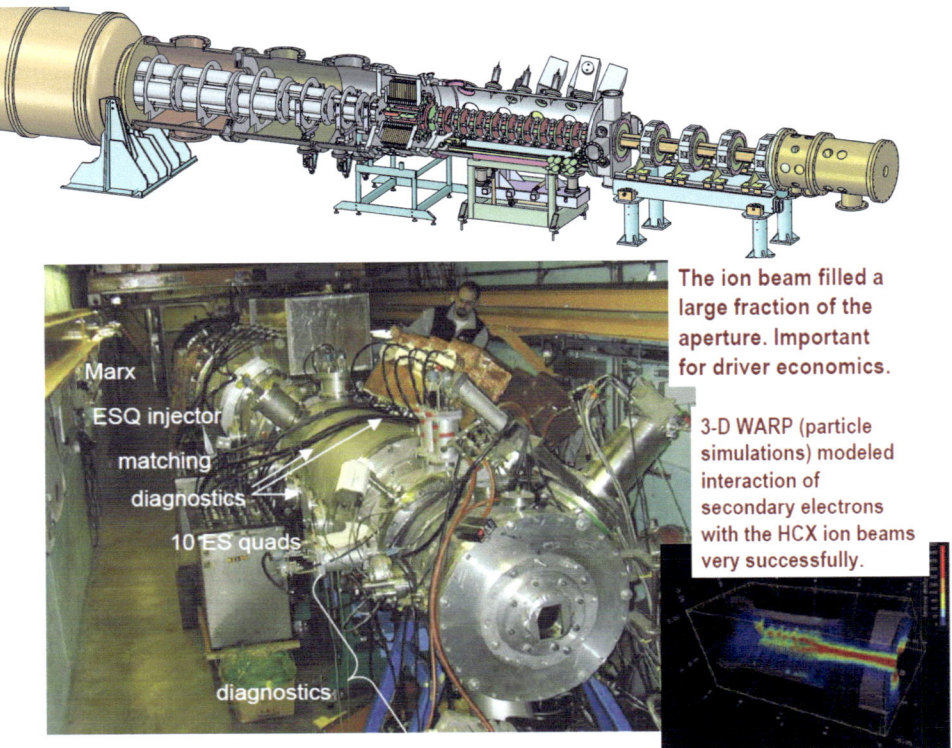

FIGURE 2.8 The HCX apparatus. SOURCE: G. Logan, LBNL, "Heavy Ion Fusion," Presentation to the committee on January 29, 2011.

- Expand the target design effort, and as NIF data come in, continually determine the implications for HIF target modeling.

Conclusion 2-9: Although no serious beam-target interaction issues have been found, the work in this area is dated. Beam parameters, particularly for some targets, have evolved into regions where the previous work may no longer be valid.

- Refine final optics design using neutronics codes; include sufficient bends to reduce the neutron flux at the end of the accelerator to a hands-on level. Assess the need for radiation-resistant plasma sources.
- Do a power plant study of the reference ≥3 MJ target approach for a liquid-wall chamber.

Medium Term (5-15 Years)

Conclusion 2-10: A very important element of the heavy ion inertial fusion energy research and development program will be the demonstration of a ≥10 kJ scale target physics facility, supporting target fabrication and injection R&D for burst-mode experiments at ~5 Hz.

This intermediate research experiment (see Chapter 4) has been proposed because, unlike the other IFE approaches, HIF has no target test bed. It is therefore critical for such an HIF facility to be able to test targets and operate in an environment as relevant to IFE as possible.

The timing for this step is discussed in Chapter 4 and Appendix J.

- Continue technology development and cost reduction with vendors for the long term.

Long Term (>15 Years)

- Construct a 2-3 MJ HIF ignition test facility, first for single-shot tests, then for burst mode, using an accelerator designed for high repetition rate. If successful, add nuclear systems to upgrade to 150 MW average-fusion-power level heavy-ion Fusion Test Facility/DEMO (HIFTF).

The programs described above are illustrated in Figure 2.9.

Observations

HIF benefits greatly from the large NNSA target physics program. The design codes are suitable for the simulation of heavy-ion targets and the target fabrication techniques are similar. Moreover, for indirect drive, the physics of the fuel capsule itself is largely independent of the source of the X-rays used to drive the fuel capsule as long as the X-rays have the correct spectrum (approximately thermal), time dependence, and symmetry.

One of the goals of the NIF is to establish the feasibility of indirectly driven targets for all drivers.[109] Although NIF can provide significant confidence in indirect drive for any driver, each driver must ultimately demonstrate that it can deliver the appropriate hohlraum conditions needed to drive the capsule.

---

[109] J.D. Lindl, P. Amendt, R.L. Berger, et al., 2004, The physics basis for ignition using indirect-drive targets on the National Ignition Facility, *Physics of Plasmas* 11: 339-491.

HIF Roadmap

```
                    2017                              2025
┌─────────────────────────────────────────────────────────────────┐
│  CHAMBER RESEARCH, NEUTRONICS, CHAMBER OPTICS INTERFACE         │
└─────────────────────────────────────────────────────────────────┘

┌────────────────────┐   ┌──────────────────────────────────────┐
│ IMPROVE:           │   │ CONTINUE ENGINEERING RESEARCH AND    │
│ ION SOURCES,       │   │ PULSING TO > 10⁹ PULSES              │
│ QUAD ARRAYS,       │   └──────────────────────────────────────┘
│ INDUCTION CORES,
│ PULSERS,                                      ┌─────────────────┐
│ INSULATORS             ┌──────────────┐       │ FUSION DRIVER   │
│                        │ INTERMEDIATE │       │                 │
│ FABRICATE:       ◇     │ TARGET       │  ◇    │ TARGET ◇ ETF ◇ DEMO │
│ INJECTOR & FRONT,      │ PHYSICS      │       │ PHYSICS         │
│ MIDDLE, AND END        │ FACILITY     │       └─────────────────┘
│ SECTIONS. RUN AT       │
│ > 10 Hz, ~10⁹ PULSES   │ DESIGN DRIVER│
└────────────────────┘   └──────────────┘
                                                ┌─────────────────┐
┌────────────────────┐                          │ TARGET          │
│ HIGH CURRENT       │                          │ FACTORY         │
│ EXPERIMENTS        │                          └─────────────────┘
└────────────────────┘

┌─────────────────────────────────────────────────────────────────┐
│ THEORY, SIMULATION    TARGET PHYSICS    EXPERIMENT, THEORY,     │
│                                         SIMULATION              │
└─────────────────────────────────────────────────────────────────┘
┌─────────────────────────────────────────────────────────────────┐
│ TARGET FABRICATION RESEARCH                                     │
└─────────────────────────────────────────────────────────────────┘
```

FIGURE 2.9 Illustrative HIF roadmap, based on the program described in the text.

Theory and existing experimental data suggest that well focused heavy-ion beams can produce the required hohlraum environment,[110] but there is currently no heavy-ion accelerator that can test the theory at the beam intensities needed for fusion. The final validation of the theory will require the construction of new facilities as shown in the roadmap above.

The heavy-ion accelerator development path differs from the development path for solid-state lasers. Much of the technology for large, solid-state lasers has been developed by the NNSA ICF program for Stockpile Stewardship. In contrast, much of the needed accelerator technology has been developed for nuclear and particle physics and, in the case of induction accelerators, for radiography and other

---

[110] See A.W. Maschke, 1975, Relativistic ions for fusion applications, Proceedings of the 1975 Particle Accelerator Conference, Washington, D.C., *IEEE Transactions on Nuclear Science* NS-22 (3): 1825; D. Eardley et al., 1983, Heavy-ion fusion, *JASON Report JSR-82-302*, The MITRE Corporation, McLean, Virginia; H.H. Heckman, H.R. Bowman, Y.J. Jarant, J.O. Rasmussen, A.I. Warwick, and Z.Z. Xu, 1987, Range energy relations for Au ions, $E/A \leq 150$ MeV, *Physical Review A* 36: 3654; D.W. Hewett, W.L. Kruer, and R.O. Bangerter, 1991, Corona plasma instabilities in heavy-ion fusion targets, *Nuclear Fusion* 31(3): 431 and references therein.

applications requiring high-current electron beams. There is an existing industrial base, but the technology must be adapted to the unique requirements of IFE.

Since accelerators are expected to be efficient and reliable and to have high pulse repetition rates, it seems possible to skip one step in the accelerator development path relative to solid-state lasers. Specifically, after building a number of smaller lasers, the laser program in the United States built two tens-of-kilojoules, single-shot laser facilities: Nova and OMEGA. The intermediate target physics facility mentioned above is of similar scale, but it is repetitively pulsed. These laser facilities were followed by the NIF. Since the NIF does not have the characteristics needed for power production, at least one additional step is required. The heavy-ion plan outlined above skips the NIF step. The proposed HIF Ignition Test Facility will initially be built without all the power supplies needed for high-repetition-rate operation. At this point, it will be used to refine and validate those aspects of target physics that have not yet been tested at full scale. The committee emphasizes again that much of the target physics, target fabrication technology, and needed diagnostics will already have been developed at the NIF and elsewhere. The final step in accelerator development program is to add the power supplies needed for high-repetition-rate operation.

Pulsed Power

Background and Status

Pulsed-power-driven IFE would utilize ≥50 MA of current from a pulsed-power accelerator to generate sufficiently high magnetic field pressures to compress and heat magnetized, preionized fusion fuel contained in a cylindrical target to ignition conditions. The pulsed-power approach has relatively low-cost and high-efficiency driver technology that appears to be scalable in a straightforward way to the peak power and total energy presently estimated to be needed for IFE. Furthermore, a high-repetition-rate technology development program is already in progress because of synergistic NNSA programs and potential commercial applications other than energy use for this technology.[111]

The primary conceptual approach to achieving pulsed-power IFE, Magnetized Liner Inertial Fusion (MagLIF), is a direct-drive approach—that is, fuel compression and heating are driven directly by magnetic pressure (see Figure 2.3). This approach offers the potential benefits of a relatively simple cylindrical target geometry and highly efficient delivery of driver energy to fuel implosion and heating. However, there is considerable uncertainty (i.e., technical risk) surrounding

---

[111] Note, however, that these commercial applications involve storing energy at much lower levels than those necessary for IFE.

all aspects of this approach owing to a paucity of relevant experimental data on target physics and ignition and a lack of in-depth design studies on inertial fusion reactors at the proposed multi-GJ yield and ~0.1 Hz repetition rate called for by the advocates. In addition to MagLIF, there other promising approaches to pulsed-power fusion energy, including one called magnetized target fusion (MTF). While MagLIF operates on the 100-ns timescale, is ~1 cm in size, and involves open magnetic field lines, MTF operates on a ~1 μs timescale, is tens of centimeters in size, and involves closed (field-reversed) magnetic field lines.

A pulsed-power fusion reactor system would be very different from both laser- and heavy-ion fusion systems. As such, the technological or economic failure modes are likely to be very different.

Historical Background

The use of <100-ns-pulse-duration, intense electron beams driven by pulsed-power generators for ICF was first discussed in the mid-1960s at Physics International Company as pulsed-power generators capable of hundreds of kiloamperes and ~10 MeV were being developed there and elsewhere.[112] F. Winterberg appears to have the earliest full publications on the subject.[113] Sandia National Laboratories (SNL) initiated a research program on pulsed-power-driven IFE with intense electron beams in the early 1970s.[114] This became the light-ion fusion program in 1979, when the advantages of intense light-ion beams relative to electrons were recognized and it became possible to produce intense light-ion beams efficiently.[115] Some progress on the generation of adequately intense light-ion beams using pulsed-power generators was made by the middle 1990s.[116] However, the demonstration of efficient coupling of electrical energy into magnetic energy and then to soft X-rays (through the intermediary of imploding cylindrical wire-array Z-pinches with hundreds of fine tungsten wires)[117] deflected the pulsed-power-driven inertial fusion community in the direction of radiation-driven (indirect-drive) fuel-capsule

---

[112] F.C. Ford, D. Martin, D. Sloan, and W. Link, 1967, *Bulletin of the American Physical Society* 12: 961.

[113] F. Winterberg, 1968, The possibility of producing dense thermonuclear plasma by an intense field emission discharge, *Physical Review* 174: 212-220.

[114] G. Yonas, J.W. Poukey, and K.R. Prestwich, 1974, Electron beam focusing and application to pulsed fusion, *Nuclear Fusion* 14: 731-740.

[115] See, for example, J.P. VanDevender, 1986, Inertial confinement fusion with light ion beams, *Plasma Physics and Controlled Fusion* 28: 841-855.

[116] J.P. Quintenz, T.A. Mehlhorn, R.G. Adams, G.O. Allshouse, et al., 1994, Progress in the light ion driven inertial confinement fusion program, *Proceedings of 15th International Conference on Plasma Physics and Controlled Nuclear Fusion Research* 3: 39-44.

[117] T.W.L. Sanford, G.O. Allshouse, B.M Marder, et al., 1996, Improved symmetry greatly increases X-ray power from wire-array Z-pinches," *Physical Review Letters* 77: 5063-5066.

implosions. The even higher potential efficiency of magnetically driven (direct-drive) ignition of magnetized fusion fuel—MagLIF—and recent favorable computer simulation results for this concept have caused MagLIF to become a leading candidate for pulsed-power fusion energy.[118]

Imploding a magnetized, field-reversed target plasma in a solid or liquid liner by a pulsed external magnetic field is a 1970s (or earlier) idea that has been pushed from the millisecond to the microsecond timescale in the present embodiment, MTF.[119] This approach is very properly described as a hybrid of magnetic and inertial confinement fusion, since the magnetic field configuration is a closed-confinement geometry. However, the duration of confinement—should fusion reactions be ignited—is determined by the inertia of the imploding liner.

Status

The necessary high-efficiency, 0.1-1 pulse-per-second pulsed-power technology is close to being in hand, and the cost per joule of energy delivered to the fusion target load is projected to be substantially lower than for all other drivers. Proof of principle that the necessary driver for a fusion reactor can be built for an acceptable price is possible within 6 years, according to the advocates.[120]

Thus far, target physics for MagLIF has been addressed only through computer simulations.[121] However, current research program plans at SNL include addressing many target physics issues using existing facilities as part of the NNSA-sponsored (single-pulse) ICF program.[122]

On the reactor side, the present MagLIF approach as proposed by SNL involves extremely high-yield pulses (~10 GJ) at a repetition rate on the order of 1 per 10 s (~0.1 Hz). This makes some of the proposed reactor challenges unique, such as the requirement for power delivery to the fusion fuel by a recyclable transmission

---

[118] M. Cuneo et al., SNL, "Pulsed Power IFE: Background, Phased R&D, and Roadmap," Presentation to the committee on April 1, 2011; M.E. Cuneo et al., SNL, Response to the committee, submitted in March 2011; S.A. Slutz, M.C. Herrmann, R.A. Vesey, et al., 2010, Pulsed-power-driven cylindrical implosions of laser pre-heated fuel magnetized with an axial magnetic field, *Physics of Plasmas* 17: 056303.

[119] G. Wurden and I. Lindemuth, LANL, "Magnetio-Inertial Fusion (Magnetized Target Fusion)," Presentation to the committee on March 31, 2011.

[120] M. Cuneo, et al., SNL, "Pulsed Power IFE: Background, Phased R&D, and Roadmap," Presentation to the committee on April 1, 2011; M.E. Cuneo et al., SNL, Response to the committee, submitted in March 2011.

[121] S.A. Slutz, M.C. Herrmann, R.A. Vesey, et al., 2010, Pulsed-power-driven cylindrical implosions of laser pre-heated fuel magnetized with an axial magnetic field, *Physics of Plasmas* 17: 056303.

[122] M. Cuneo et al., SNL, "Pulsed Power IFE: Background, Phased R&D, and Roadmap," Presentation to the committee on April 1, 2011; M.E. Cuneo et al., SNL, Response to the committee, submitted in March 2011.

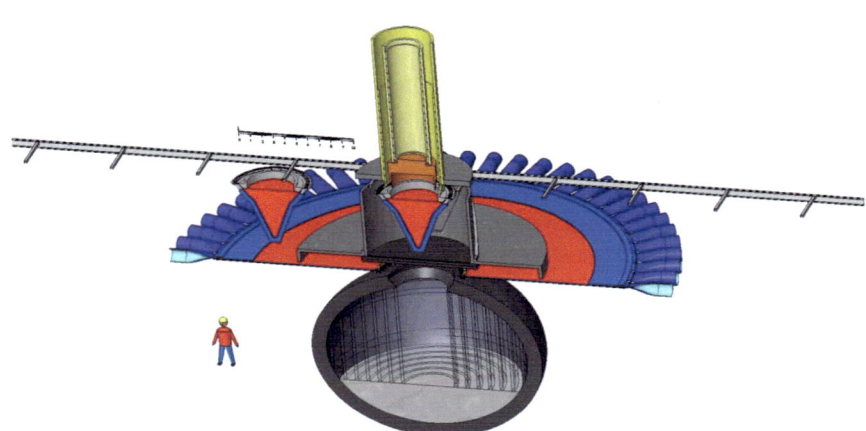

FIGURE 2.10 Recyclable transmission line concept with liquid wall chamber. SOURCE: M. Cuneo, SNL, Presentation to the committee on April 1, 2011.

line (RTL) (see Figure 2.10).[123,124] There has been some analysis, and some small-scale experiments have been carried out that address how such high yields might be sustained repetitively in a reactor chamber.[125]

Single-pulse tests of MTF are being done now with the Shiva Star facility at the Air Force Research Laboratory at 6 MA. Next-generation tests are proposed that would use explosively driven, high-magnetic-field generation to drive the implosion, but IFE would require a high-repetition-rate pulsed-power driver. Reactor considerations for this concept have not been developed in detail to the committee's knowledge.

---

[123] The recyclable transmission line is destroyed during each shot. Because it contains a considerable mass of material, economical operation dictates that this material be recycled.

[124] See M. Cuneo et al., SNL, "Pulsed Power IFE: Background, Phased R&D, and Roadmap," Presentation to the committee on April 1, 2011; M.E. Cuneo et al., SNL, Response to the committee, submitted in March 2011; and J.T. Cook, G.E. Rochau, B.B. Cipiti, et al., Z-inertial fusion energy: Power plant final report FY06, Sandia National Laboratories report SAND2006-7148.

[125] See J.T. Cook, G.E. Rochau, B.B. Cipiti, C.W. Morrow, S.B. Rodriguez, C.O. Farnum, et al., 2006, Z-inertial fusion energy: Power plant, SAND2006-7148, Sandia; M. Sawan, L. El-Guebaly, and P. Wilson, 2007, Three dimensional nuclear assessment for the chamber of Z-pinch power plant, *Fusion Science and Technology* 52: 753; S.B. Rodríguez, V.J. Dandini, V.L. Vigíl, and M. Turgeon, 2005, Z-pinch power plant shock mitigation experiments, modeling and code assessment, *Fusion Science and Technology* 47: 656; S.I. Abdel-Khalik and M. Yoda, 2005, An overview of Georgia Tech studies on the fluid dynamics aspects of liquid protection schemes for fusion reactors, *Fusion Science and Technology* 47: 601; S.G. Durbin, M. Yoda, and S.I. Abdel-Khalik, 2005, Flow conditioning design in thick liquid protection, *Fusion Science and Technology* 47: 724.

Scientific and Engineering Challenges and Future R&D Priorities for Pulsed-Power IFE Applications

Implosion of magnetized plasma inside a conducting cylinder on open field lines to achieve fusion ignition depends on magnetic inhibition of radial energy transport and effective fusion burn before the hot plasma can run out the ends. MagLIF would achieve this with a ~100 ns implosion time and a few centimeters of high-density plasma confined by open magnetic field lines. Thus, the major target physics challenges that are to be addressed in the near term on Z are the following:

- Demonstrating that the predicted high-efficiency energy transfer from electrical energy to hot magnetized fusion fuel plasma compressed by magnetic-field-driven implosion of a cylindrical conducting liner occurs in experiments. Determining plasma conditions inside the imploding liner is a major part of this challenge.
- Demonstrating that the energy-loss rate of the compressed plasma is much less than that of an unmagnetized plasma. Understanding how the magnetic field affects the transport coefficients is a necessary part of this research to allow validating the design codes.

The MTF version of the two items is to demonstrate at 6 MA that a sufficiently well-confined plasma can be produced to warrant explosively driven experiments that have a much higher cost than the pulsed-power experiments. As in MagLIF, diagnostic access to the plasma if it is not generating the predicted number of neutrons is very limited, again making determination of the plasma condition inside the liner a part of this challenge.

The biggest early technology challenge for pulsed-power IFE is establishing the technical credibility of the proposed low-repetition-rate (~0.1 Hz), ~10 GJ yield-per-pulse reactor concept. The recyclable transmission line approach for delivering the current from the pulsed-power system to the fusion-fuel-containing target must be demonstrated to be technically feasible. Technical issues that must be addressed for the transmission line include these: what material to use, how thick it must be, and how to recycle it economically; how best to load the assembly in the reactor chamber (bearing in mind that the fusion-fuel-containing load—possibly requiring cryogenics—must be attached to it); and how to assure that the assembly makes a good electrical connection to the pulsed-power system.

Demonstrating the engineering feasibility of a thick-liquid-wall reactor chamber is a challenge that pulsed-power shares with other possible approaches, particularly heavy-ion fusion. However, pulsed-power fusion, as most recently proposed, is alone in requiring compatibility of the reactor chamber with recyclable transmission lines and with ~10 GJ yield per pulse (the equivalent of 2.5 tons of high explosive). Some analyses of fatigue and nucleonics limits of possible chamber materials and some experimental studies relevant to thick liquid wall reactor chambers have been carried out,[126] but much work is yet to be done here. Design and execution of a hydrodynamically equivalent experiment that could be conducted in a smaller "scaled" chamber at a much-reduced energy level should be part of the Phase 1 research program (see Table 2-3). This research would benefit heavy-ion fusion as well. If there is no technically viable solution to the reactor chamber problem at 10 GJ that is also economically viable, then pulsed-power fusion researchers will have to reoptimize their system design at a lower energy per pulse and a higher repetition rate than 0.1 Hz. Thus, the technical and economic feasibility of the 10 GJ yield system should be evaluated as early in Phase 1 as possible.

Given the state of development of linear transformer drivers (LTDs) (see Figures 2.11 and 2.12),[127] the technology challenges associated with the pulsed-power system appear to be much less daunting than those discussed above. Nevertheless, the technology must still be demonstrated to be extremely reliable, as there would be hundreds of thousands of switches and a million capacitors in a pulsed-power reactor driver.[128] Furthermore, the driver must be demonstrated to be compatible with using recyclable transmission lines, including their potential failure modes (e.g., sparking due to poor connections).

Many of the scientific issues having to do with MagLIF target physics can be addressed using existing facilities in the next 5 years, and many will be investigated as part of the NNSA-sponsored (single-pulse) ICF program at SNL. It is anticipated that this program will be funded at an estimated level of $6.8 million to $8.5 million per year through 2017.[129] All pulsed-power approaches call for recyclable transmission lines and extremely high-yield pulses at a repetition rate of ~0.1 Hz, and these requirements make some of the necessary research and development for pulsed-power IFE unique. The high repetition rate driver technology needed for fusion via pulsed power is currently receiving development funding at the rate of $1.5 million to $3.3 million per year,[130] and steady progress is being made.

The engineering feasibility challenges of MagLIF should be addressed early in the program, along with the target physics, to assess the viability of pulsed-power fusion. To do this, new funding would be required starting in 2013 at the level

---

[126] Ibid.

[127] W. Stygar, SNL, "Conceptual Design of Pulsed Power Accelerators for Inertial Fusion Energy," Presentation to the committee on April 1, 2011.

[128] J.T. Cook, G.E. Rochau, B.B. Cipiti, C.W. Morrow, S.B. Rodriguez, C.O. Farnum, et al., 2006, Z-inertial fusion energy: Power plant, SAND2006-7148.

[129] M. Cuneo, personal communication to committee member D. Hammer on November 2, 2011.

[130] Ibid.

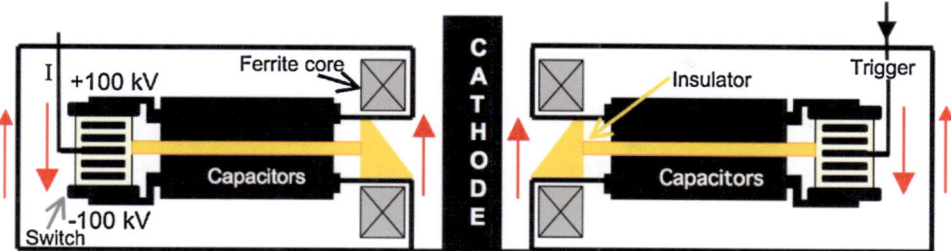

FIGURE 2-11 Pictorial representation of a side section of an annular LTD cavity where the load now is the coaxial line formed by the inner cylindrical surface of the cavity and the central (cathode) cylindrical electrode. The red arrows show the current direction in each conductor. Each unit consists of two capacitors charged to ±100 kV, a 200-kV switch, and a portion of the annular ferrite cores that assure that the pulse is delivered to the load until the cores saturate. There are many such units in parallel around the annular cavity in order to produce the desired output current. SOURCE: Copied with permission of the first author from M.G. Mazarakis, W.E. Fowler, A.A. Kim, et al., 2009, High current, 0.5-MA, fast, 100-ns, linear transformer driver experiments, *Physical Review Special Topics-Accelerators and Beams* 12: 050401.

FIGURE 2-12 Top view of 20 units in parallel in an annular LTD cavity. SOURCE: Copied with permission of the first author from M.G. Mazarakis, W.E. Fowler, A.A. Kim, et al., 2009, High current, 0.5-MA, fast, 100-ns, linear transformer driver experiments, *Physical Review Special Topics-Accelerators and Beams* 12: 050401.

TABLE 2.3 Elements of a Pulsed-Power Inertial Fusion Energy Program

Phase 1	Phase 2	Phase 3 (FTF)
MagLif target physics	Target physics: achieve ignition on a single-pulse facility with repetition-rate-capable pulsed-power technology.	Build and test an FTF that operates in burst mode and is capable of achieving breakeven.
Validate codes		
LTD technology development		Achieve multigigajoule yield per pulse.
RTL engineering studies	Establish the viability of a 0.1 Hz, 10-GJ-yield IFE facility through analysis and scaled hydrodynamics experiments.	
Reactor chamber engineering studies		
Infrastructure planning (targets, etc.)	Demonstrate RTL engineering feasibility in burst mode.	
	Design an FTF for pulsed-power IFE.	

of $8 million to $10 million per year if a Technology Readiness Level of 6 (see Chapter 4) by 2018 is to be achieved for many of the elements of the reactor.[131]

> **Conclusion 2-11:** The promise of MagLIF as a high-efficiency approach to inertial confinement fusion is largely untested, but the program to do so is in place and is funded by NNSA.
>
> **Conclusion 2-12:** There has been considerable progress in the development of efficient pulsed-power drivers of the type needed for inertial confinement fusion applications, and the funding is in place to continue along that path.
>
> **Conclusion 2-13:** The physics challenges associated with achieving ignition with pulsed power are being addressed at present as part of the NNSA-sponsored (single-pulse) inertial confinement fusion program.
>
> **Recommendation 2-2:** Physics issues associated with the Magnetized Liner Inertial Fusion (MagLIF) concept should be addressed in single-pulse mode during the next 5 years so as to determine its scientific feasibility.

---

[131] M. Cuneo et al., SNL, "Pulsed Power IFE: Background, Phased R&D, and Roadmap," Presentation to the committee on April 1, 2011, and M.E. Cuneo et al., SNL, Response to the committee, submitted in March 2011.

Conclusion 2-14: The major technology issues that would have to be resolved in order to make a pulsed-power IFE system feasible—the recyclable transmission line and the ultra-high-yield chamber technology development—are not receiving any significant attention.

Recommendation 2-3: Technical issues associated with the viability of recyclable transmission lines and 0.1 Hz, 10-GJ-yield chambers should be addressed with engineering feasibility studies in the next 5 years in order to assess the technical feasibility of MagLIF as an inertial fusion energy system option.

Assuming the necessary milestones are achieved in both target physics and engineering feasibility, a second phase that would last about 10 more years could be undertaken starting around 2018 to develop the necessary reactor-scale technology and industrial capacity for an FTF.

Some of the necessary technology infrastructure—specifically, production of the recyclable transmission line—may be close enough to standard large-scale industrial manufacturing that development costs and schedule can be projected with reasonable confidence without major demonstration projects. The fact that the cylindrical fusion fuel-containing targets for MagLIF will be inserted into the reactor chamber as part of the recyclable transmission line assembly is a potential simplification compared to other IFE approaches, assuming viable engineering solutions for the line's fabrication, emplacement, contact, and recycling problems are found.

The MTF has a 3-year target physics program plan using Shiva Star at $2.8 million per year, which is to be followed by explosively driven implosion tests in Nevada at about $100 million per year for 2 years.

Path Forward for Pulsed-Power Inertial Fusion Energy

The plan for pulsed-power IFE that follows is based on information provided to the committee by SNL.

Near Term (≤5 Years, Initially Using NNSA Funding)

- *Target physics.* Using existing facilities, validate the magnetically imploded cylindrical target concept to the point of achieving scientific breakeven (fusion energy out = energy delivered to the fuel). This requires developing tritium-handling capability on Z. Also, develop IFE target requirements experimentally and theoretically, which requires validating computer codes.
- *Pulsed power.* Demonstrate the capability of Linear Transformer Driver pulsed-power technology to deliver the necessary power, energy, and

repetition rate with a long operational lifetime and the anticipated high efficiency. Design the reactor driver.
- *Recyclable transmission line.* Develop an engineering design of a recyclable (magnetically insulated) transmission line (RTL) and demonstrate its engineering feasibility experimentally at high power (low repetition rate).
- *Reactor chamber.* Carry out a detailed design study of the presently favored, multigigajoule, thick-liquid-wall, low-repetition-rate (~0.1 Hz) reactor concept; develop the conceptual design of a credible demonstration power plant in partnership with industry; initiate necessary technology R&D. Design and, if warranted, implement a hydrodynamically equivalent test of the viability of a thick-liquid-wall chamber to contain repeated 10 GJ yield fusion explosions. Determine with industrial partners if such a low-repetition-rate, high-yield system is the optimum solution for pulsed power in light of target physics, recyclable transmission line, and pulsed-power ICF/IFE developments in Phase 1 (Table 2.3).
- *Industrial infrastructure planning.* In partnership with industry, design production lines and delivery systems needed for RTLs, targets, etc.
- *Next facility design.* Determine the necessary new facility for ignition experiments (defined as fusion alpha-particle heating of the fuel exceeding energy delivered to the fuel by the driver) and high yield (up to 100 MJ), from which the fusion burn can be scaled to the ~10 GJ yield per target needed by the reactor. (See ZFIRE in the pulsed-power IFE roadmap in Figure 2.13.)

New funding between $8 million and $10 million per year is needed to undertake the last four engineering development tasks.[132]

Medium Term (5-15 Years, Assumes All Milestones in Phase 1 Are Achieved)

- *Target physics: Ignition.* Achieve ignition in a new, repetitive-pulse-capable Linear Transformer Driver pulsed-power facility (ZFIRE); fully validate design codes needed to scale to full reactor yield. This would be an NNSA facility that can be used for weapon physics and weapon effects testing.
- *Recyclable transmission line engineering.* Demonstrate operation of an RTL at ~100 TW and 0.1 Hz (burst mode), with ignition for one or more "single pulses."
- *Reactor chamber.* Establish by analysis and demonstrate key technologies associated with the thick liquid wall IFE reactor chamber needed for ~10 GJ, 0.1 Hz operation (vacuum system, liquid wall recovery, and so on). This technology could also be beneficial for heavy-ion fusion.

---

[132] M. Cuneo et al., op. cit.

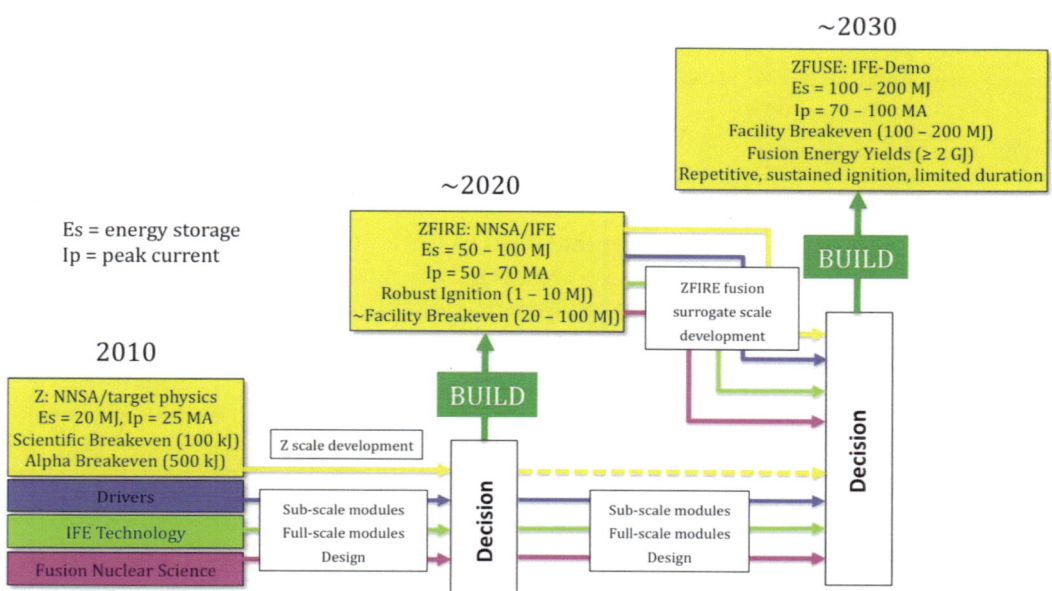

FIGURE 2.13 Pulsed-power roadmap. SOURCE: M.E. Cuneo, M.C. Herrmann, W.A. Stygar, et al., from the document submitted to the committee in response to the committee's Second Request for Input, p. 6, received March 24, 2011.

- *Target design and fabrication for inertial fusion energy.* Determine optimized target design and target fabrication requirements for an FTF and a demonstration power plant.
- *Fusion Test Facility design.* With industry, develop an engineering design for an FTF for pulsed-power fusion, including factories to build RTLs, targets, and other components that must be replaced with each pulse; tritium breeding and handling systems; and all balance-of-plant systems. Design must include full resource requirement and safety and reliability analyses. An economically "competitive" cost of electricity must be projected or this approach cannot go to the demonstration stage.

There are two aspects to the cost of electricity: the amortized capital cost of the plant, the estimate of which is likely to be better than a factor of two only at the end of Phase 2 (see Table 2.3), and the cost of plant operation. Included in the latter is fuel cost, including operation of the tritium recovery system. Let us assume that is the same for all of the potential reactors. The dominant additional operating cost for pulsed-power fusion energy is likely to be manufacturing and recycling the RTLs. At present it is not known how that will compare with, for example, the actual costs incurred by laser-driven systems for replacing optical components or heavy-ion fusion for replacing final focusing magnets. This kind of operating cost will not be known very well until the end of Phase 2 for any of the approaches to IFE.

Long Term (>20 Years): Build and Operate a Fusion Test Facility

Assuming all milestones in the medium-term program are met, an FTF would be designed to achieve facility breakeven in initial operation (fusion yield of 100-200 MJ) in repetitive pulse operation but for "bursts" of limited duration. Upgrades would enable this facility to increase its yield to ~2 GJ or more. It is too early to provide a credible estimate for the cost of an FTF (see ZFUSE in the Roadmap, below) as the cost of the reactor chamber and recyclable transmission line factory are likely to be dominant and they will not be established until the end of Phase 2.

A conceptual roadmap for implementing the R&D program for pulsed power inertial fusion is shown in Figure 2.13.

GENERAL CONCLUSIONS

There are a number of technical approaches, each involving a different combination of driver, target, and chamber, that show promise for leading to a viable IFE power plant. These approaches involve three kinds of targets: indirect drive, direct drive, and magnetized target. In addition, the chamber may have a solid or a thick-liquid first wall that faces the fusion fuel explosion, as discussed in Chapter 3.

Substantial progress has been made in the last 10 years in advancing most of the elements of these approaches, despite erratic funding for some programs. Nonetheless, substantial amount of R&D will be required to show that any particular combination of driver, target, and chamber would meet the requirements for a demonstration power plant.

In all cases, the drivers may build on decades of research in their area. In all technical approaches there is the need to build a reactor-scale driver module for use in an FTF. The timing for this step is discussed in Chapter 4.

As discussed in Chapter 4, development of an FTF and the upgrade to a demonstration plant requires an integrated system engineering approach supported by R&D at each stage. This statement is true regardless of which driver-target combination is chosen. It also requires involvement and support from the user community (utilities), from the facilities engineering community (large engineering firms), and the government (national laboratories) to conduct R&D and risk reduction programs for laser drivers, target physics, target manufacturing and commissioning, reactors, and balance-of-plant systems. In addition, work must address licensing and environmental and safety issues.

3

Inertial Fusion Energy Technologies

This chapter deals with the technologies other than the driver technologies covered in Chapter 2 that are required to produce and utilize the energy from fusion nuclear reactions in an inertial fusion energy (IFE) system. The first sections in this chapter cover the targets, chambers, related materials issues, as well as tritium production and recovery. Subsequent sections cover crosscutting issues of environment, health, and safety as well as balance-of-plant and economic considerations.

In addition to target science, there are challenges for IFE embedded in what is usually labeled "technology" (e.g., chambers): These challenges involve a broad range of scientific disciplines, including nuclear and atomic physics, materials and surface science, and many aspects of engineering science. In the next several years, however, IFE research will be involved not in engineering developments, but rather in science and engineering research aimed at determining whether feasible solutions exist to very challenging "technology" problems.

An effort is needed to determine whether there is any IFE concept (where "concept" means some combination of target type, driver, and chamber) that appears to be feasible. Only certain combinations of targets, drivers, and chambers seem to be workable. While the emphasis today and in the near future should be on target performance, working exclusively on problems associated with target performance could easily lead to solutions that are not compatible with practical driver and chamber options. Such a serial approach could lead to dead ends and could also extend the time it takes to arrive at practical applications of IFE. For each technological approach, the committee identifies a series of critical R&D objectives that must be met for that approach to be viable. If these objectives cannot be met, then other approaches will need to be considered.

The approach used in the High Average Power Laser (HAPL) program (see Chapter 1) was one in which all the potential feasibility issues of the entire IFE system were studied, and then the most important ones were addressed to try to find basic solutions. It is a good example of how a national IFE program might be structured.

HIGH-LEVEL CONCLUSIONS AND RECOMMENDATIONS

The main high-level conclusions and recommendations from this chapter are given below.

Conclusions

Conclusion 3-1: Technology issues—for example, chamber materials damage and target fabrication and injection—can have major impacts on the basic feasibility and attractiveness of IFE and thus on the direction of IFE development.

Conclusion 3-2: At this time, there appear to be no insurmountable fusion technology barriers to realizing the components of an IFE system, although knowledge gaps and large performance uncertainties remain, including those surrounding the performance of the system as a whole.

Conclusion 3-3: Significant IFE technology research and engineering efforts are required to identify and develop solutions for critical technology issues and systems such as targets and target systems; reaction chambers (first wall/blanket/shield); materials development; tritium production, recovery and management systems; environment and safety protection systems; and economic analysis.

Recommendations

Recommendation 3-1: Fusion technology development should be an important part of a national IFE program to supplement research in IFE science and engineering.

Recommendation 3-2: The national inertial fusion energy technology effort should leverage materials and technology development from magnetic fusion energy efforts in the United States and abroad. Examples include ITER's test blanket module R&D program, materials development, plasma-facing components, tritium fuel cycle, remote handling, and fusion safety analysis tools.

TARGET FABRICATION AND HANDLING FOR INERTIAL FUSION ENERGY

Sufficiently rapid fabrication of targets that meet the exacting specifications needed to achieve high gain and an acceptable cost has long been recognized as a key requirement of practical energy application of inertial fusion. All of the earlier National Research Council (NRC) studies on IFE commented on the importance of target fabrication to the success of inertial fusion for energy applications and noted that the prospects for success appear favorable albeit with much work remaining.[1] Most of the many IFE power plant design studies have given serious consideration to how the target fabrication requirements could be achieved.[2] The consensus of these studies is that with adoption of a limited number of target designs, the selection of mass fabrication techniques, and a development program, the required accuracy and cost goals may be achieved. The R&D needed to make these projections a reality has begun with efforts at General Atomics, the Lawrence Livermore National Laboratory (LLNL) and the University of Rochester. This recent work has focused primarily on laser-driven targets, both direct and indirect drive. Earlier work on ion-beam-driven targets indicates that similar conclusions are expected to hold. Pulsed-power target development is at an early stage, but the

---

[1] "Summary of the Findings and Recommendations of the 1986, 1990, and 1997 National Research Council's Reviews of the Department of Energy's Inertial Confinement Fusion Program," Document prepared by NRC staff member E.E. Boyd and provided to the committee on March 2, 2011.

[2] For example, see the following: D.T. Goodin, N.B. Alexander, L.C. Brown, et.al., 2005, Demonstrating a target supply for inertial fusion energy, *Fusion Science and Technology* 47: 1131-1138; D.T. Frey, N.B. Alexander, A.S. Bozek, D.T. Goodin, R.W. Stemke, T.J. Drake, and D. Bitner, 2007, Mass production methods for fabrication of inertial fusion targets, *Fusion Science and Technology* 51: 786-790; L.R. Foreman, P. Gobby, J. Bartos, et al., 1994, Hohlraum manufacture for inertial confinement fusion, *Fusion Technology* 26: 696-701; M.J. Monsler and W.R. Meier, 1994, Automated target production for inertial fusion energy, *Fusion Technology* 26: 873-880; K.D. Wise, T.N. Jackson, N.A. Masnari, et al., 1979, A method for the mass-production of ICF targets, *Journal of Nuclear Materials* 85-86: 103-106; B.A. Vermillion, J.T. Bousquet, R.E. Andrews, et al., 2007, Development of a new horizontal rotary GDP coater enabling increased production, *Fusion Science and Technology* 51: 791-794; J.T. Bousquet, J.F. Hund, D.T. Goodin, and N.B. Alexander, 2009, Advancements in glow discharge polymer coatings for mass production, *Fusion Science and Technology* 55: 446-449; W.S. Rickman and D.T. Goodin, 2003, Cost modeling for fabrication of direct drive inertial fusion energy targets, *Fusion Science and Technology* 43: 353-358; K.R. Schultz, 1998, Cost effective steps to fusion power: IFE target fabrication, injection and tracking, *Journal of Fusion Energy* 17: 237-246.

slower repetition rate (~0.1 Hz as opposed to 10 Hz) and the simple target design should ease the challenges of target fabrication for pulsed power. However, much remains to be done for IFE target development for all drivers.

The committee concurs with the conclusion that suitable target fabrication will be possible at an acceptable cost, so that target fabrication is not an obviously insurmountable obstacle for IFE. However, the committee does not endorse the projected target cost numbers any more than it endorses estimates of future costs for any component of IFE technology in the early development stage. The costs could be much higher or lower than estimated in the conceptual studies that have been done. Only a substantial national development effort will provide the validation needed.

When and if ignition is reached, it will be necessary to turn more attention and devote greater resources to target fabrication development. Concepts for producing targets at a rate 100,000 times the rate at which targets are produced today have been developed; therefore, if and when ignition is reached, it would be a good time to determine if the target factory components can be validated with real equipment and if a small, complete factory operating at modest production rates can be built and operated successfully. Such a facility should be accompanied by continued development, begun under the inertial confinement fusion (ICF) program, of physics models of the formation of small hollow spheres, subsequent deuterium-tritium (DT) layering, and other fabrication processes.

Background and Status[3]

For direct drive, an inertial fusion target consists of a spherical capsule that contains a smooth layer of DT fuel. For indirect drive, the capsule is contained within a metal "hohlraum" that converts the driver energy into X-rays to drive the capsule. These concepts are shown schematically in Figure 3.1. For pulsed-power, target designs vary from those similar to indirect drive, to cylindrical metal shells containing DT. Several examples of IFE targets are shown in Figure 3.2.

Fusion fuel targets must be delivered in a form that meets the stringent requirements of the particular inertial fusion energy scheme, in sufficient quantity and at a low enough cost to supply affordable electricity to the grid. A fusion power plant will consume as many as 1 million targets per day. The allowable target cost will depend on the maximum marketable cost of electricity and the target yield, with estimates for laser and heavy-ion beam systems of 20-40 cents each, based on conceptual modeling studies. For higher-yield, pulsed-power systems, the cost could be proportionately higher. The cost of raw materials for the targets under

---

[3] Portions of this discussion are taken from Appendix C of the Department of Energy's (DOE's) Fusion Energy Sciences Advisory Committee 1999 report *Summary of Opportunities in the Fusion Energy Sciences Program*.

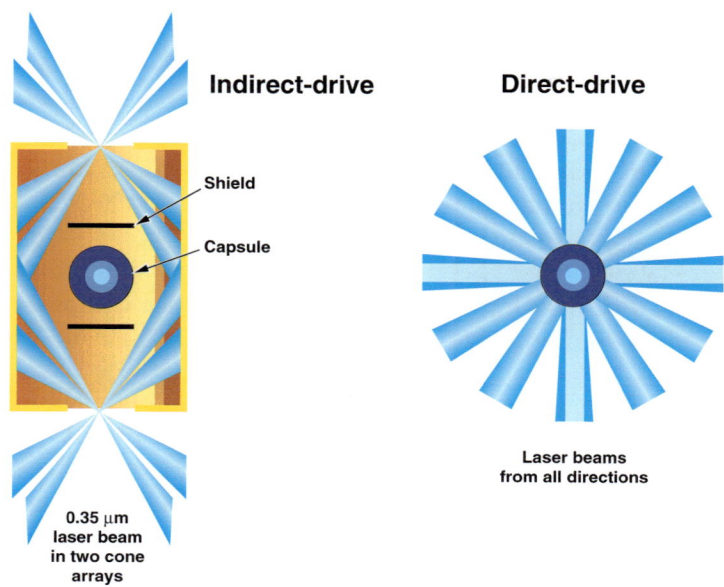

FIGURE 3.1 Indirect-drive and direct-drive IFE target concepts. SOURCE: LLNL.

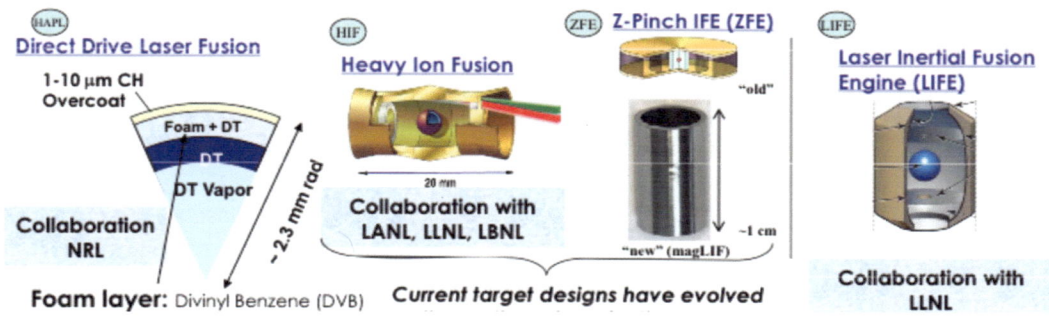

FIGURE 3.2 Examples of IFE targets used with various driver schemes. NRL, Naval Research Laboratory; LANL, Los Alamos National Laboratory, LLNL, Lawrence Livermore National Laboratory, LBNL, Lawrence Berkeley National Laboratory. SOURCE: General Atomics.

consideration currently is at the few-cents-per-target level. Mass manufacturing experience in other industries suggests that these production cost goals are possible, but a development program is required to validate the conceptual modeling studies. Current target production costs and rates are not useful for estimating the costs of mass-produced targets, although the gap between what can be done today and what is needed indicates that target fabrication for IFE plants is a challenge.

The fabrication techniques currently used for inertial confinement fusion (ICF) research targets must meet exacting specifications, have maximum flexibility to accommodate changes in target designs, and provide thorough characterization for each target. Current ICF target fabrication techniques for research targets may not be well suited to economical mass production of IFE targets. Because of the large number of designs and the thorough characterization required for each target, an ICF research target can currently cost thousands of dollars apiece. However, IFE target mass-fabrication studies are encouraging. Fabrication techniques are proposed that are well suited for economic mass production and promise the precision, reliability, and economy needed. However, work has just begun on these techniques.

- *Fuel capsules.* The capsules must meet stringent specifications including out-of-round ($d_{max} - d_{min} < 1$ μm), wall thickness uniformity ($\Delta w < 0.5$ μm), and surface smoothness (<200 Å rms).[4] The microencapsulation process, by which tiny particles or droplets are surrounded by a coating, appears well-suited to IFE target production if sphericity and uniformity can be maintained as the capsule size is increased from current 0.5- to 2-mm capsules to the ~5-mm-diameter capsule needed for IFE. Microencapsulation also appears to be suited to the production of foam shells, which are needed for several IFE target designs. Capsule designs for OMEGA experiments and direct-drive IFE power plants are shown in Figure 3.3.
- *Hohlraums.* ICF hohlraums are currently made by electroplating the hohlraum material, generally gold, onto a mandrel that is then dissolved, leaving the empty hohlraum shell. This technique does not scale up for mass production. Stamping, die-casting, and injection molding, however, do hold promise for IFE hohlraum production.[5]
- *Target assembly.* ICF research targets are currently assembled manually using micromanipulators under a microscope. Placement of the capsule at the center of the hohlraum must be accurate to within 25 μm. For IFE, this process must be fully automated, which appears possible. Initial efforts with robotic target assembly and snap-together alignment techniques have shown promising results.[6]
- *Target characterization.* Precise target characterization of every research target is needed to prepare the complete "pedigree" required by the ICF experimentalists. Characterization for current research targets is largely

---

[4] D. Goodin, General Atomics, "Target Fabrication and Injection Challenges in Developing an IFE Reactor," Presentation to the committee on January 30, 2011.

[5] A. Nikroo, General Atomics, "Technical Feasibility of Target Manufacturing," Presentation to the committee on July 7, 2011.

[6] A. Nikroo, during a site visit to General Atomics on February 22, 2012.

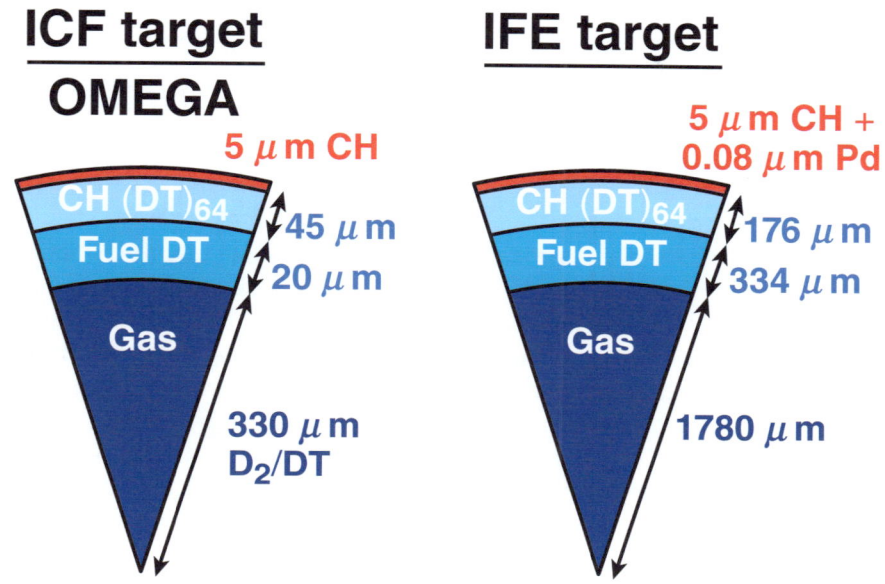

FIGURE 3.3 Direct-drive target capsules. SOURCE: University of Rochester.

done manually and is laborious. For IFE the target production processes must be sufficiently repeatable and accurate that characterization can be fully automated and used only with statistical sampling of key parameters for process control.

- *DT filling and layering.* Targets for ICF experiments are filled by permeation, and a uniform DT ice layer is formed by "beta layering." Using very precise temperature control, excellent layer thickness uniformity and surface smoothness of about 1 μm rms can be achieved.[7] These processes are suited to IFE, although the long fill and layering times needed may result in large (up to ~10 kg) tritium inventories. Advanced techniques, such as liquid wicking into a foam shell, could greatly reduce this amount. These processes are improving but remain far short of the level of reproducibility that a reactor would require. If IFE targets need DT ice smoothness of better than ~1 μm to achieve high gain, new layering techniques will be needed.
- *Target handling and injection.* IFE targets will be injected into the target chamber at rates as high as ~10-20 Hz. The targets must have adequate

---

[7] D. Goodin, General Atomics, "Target Fabrication and Injection Challenges in Developing an IFE Reactor," Presentation to the committee on January 30, 2011.

thermal and mechanical robustness and protection, such as hohlraums or sabots, to survive the injection and in-chamber flight. This solution must also be compatible with the chamber protection and energy recovery schemes (see the next section, "Scientific and Engineering Challenges and R&D Priorities.").

ICF research targets that meet all current specifications for both laser direct and indirect drive have been fabricated and fielded in small quantities, including the uniform, smooth DT ice layer. ICF research targets currently cost thousands of dollars apiece on average, but the costs vary widely; simple production targets can cost many times less, and targets requiring significant development effort could cost many times more than that amount. For a power plant, a significant transition needs to be undertaken using low-cost, high-throughput manufacturing techniques, along with large batch sizes for any chemical processes, as well as likely use of statistical characterization. Many of the processes used for current target fabrication do not scale well to mass production and will need to be replaced. Examples are die-casting arrays of hohlraum parts instead of diamond turning a mandrel for gold plating, and the use of large-batch chemical vapor deposition (CVD) diamond coaters for the ablators and membranes instead of the small size bounce-pan coaters now used. The HAPL program, led by the Naval Research Laboratory (NRL), which went well beyond laser drivers to consider all aspects of IFE power by laser direct drive, and the Laser Inertial Fusion Energy (LIFE) program, led by LLNL, which focuses on IFE by laser indirect drive, have begun evaluation and selection of mass production methods that can meet IFE requirements. The termination of the HAPL program has slowed this effort.

There have been successful efforts to develop several IFE target mass production techniques. To make thick-walled polymer capsules, a poly-alpha-methyl-styrene (PAMS) mandrel is made by microencapsulation and then coated with glow discharge polymer (GDP). A rotary kiln version of the GDP coater has been made that is capable of mass production, but it has not been used enough to demonstrate that it can meet the surface roughness specification.[8] In the HAPL program,[9] foam shells were made that met the HAPL target specification with appreciable yield using microencapsulation droplet generators. Applying a smooth, gastight overcoat to these foam shells was the focus of development at the time that the HAPL program ended. A cryogenic fluidized bed for layering deuterium in direct-drive targets was built in the HAPL program. It was successfully operated at cryogenic

---

[8] A. Nikroo, General Atomics, "Technical Feasibility of Target Manufacturing," Presentation to the committee on July 7, 2011.

[9] J.D. Sethian, D.G. Colombant, J.L. Giuliani, et al., 2010, The science and technologies for fusion energy with lasers and direct-drive targets, *IEEE Transactions on Plasma Science* 38 (4): 690-703.

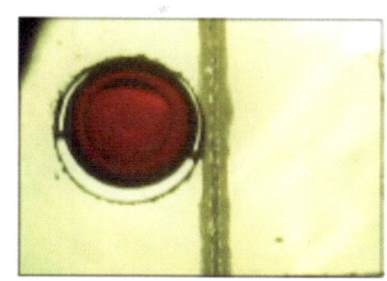

FIGURE 3.4 Electric-field-mediated microfluidics ("lab-on-a-chip") wicking of cryogenic D_2 into a foam capsule target. ITO, indium tin oxide. SOURCE: University of Rochester.

temperatures using empty capsules but has yet to be operated with deuterium-filled capsules. General Atomics has built a robotic target assembly station based on commercially available industrial robots. This station has glued together cone-in-shell targets suitable for fast ignition experiments[10] such that the virtual cone tip coincides with the capsule center to within the specification of 10 µm. LLNL is developing target assembly techniques for the National Ignition Facility's (NIF's) National Ignition Campaign (NIC) that facilitate target component self-alignment ("snap-together" assembly), which will be useful for IFE target assembly. Development of the process for manufacturing hohlraum parts made of lead by cold forging (or stamping) started recently. Some development of die-casting hohlraum parts is also expected to begin soon.[11] Innovative concepts such as the University of Rochester's use of electric-field mediated microfluidics (lab-on-a-chip),[12] shown in Figure 3.4, may allow higher quality at lower cost. In summary, progress has been made on IFE target fabrication, creating many opportunities for improved materials and technologies, but much remains to be done.

---

[10] A. Nikroo, during a site visit to General Atomics on February 22, 2012.

[11] A. Nikroo, General Atomics, "Technical Feasibility of Target Manufacturing," Presentation to the committee on July 7, 2011.

[12] D.R. Harding, T.B. Jones, Z. Bei, W. Wang, S.H. Chen, R.Q. Gram, M. Moynihan, and G. Randall, 2010, *Microfluidic Methods for Producing Millimeter-Size Fuel Capsules for Inertial Fusion*, Materials Research Society Fall Meeting, Boston, Mass.

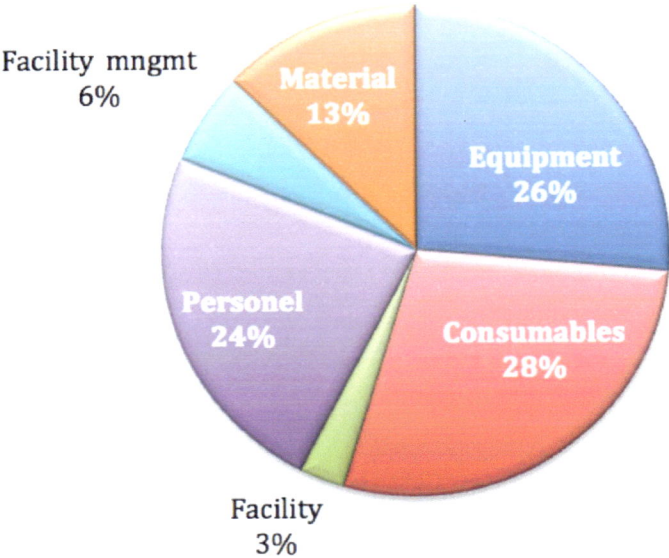

FIGURE 3.5 Cost breakout for target mass manufacture, based on a representative factory model (example shown for LIFE targets). SOURCE: R. Miles, J. Biener, S. Kucheyev, et al., 2008, "LIFE Target Fabrication Research Plan," LLNL-TR-408722.

To estimate possible costs, factory models have been constructed based on experience from the chemical batch processing industry combined with in-house expertise at General Atomics and LLNL. These models considered likely manufacturing and assembly equipment types, factory build costs, personnel and operational costs, and in-process volumes, among other things, and amortized the integrated costs over the volume of targets produced. Predictions ranged from 17 to 35 cents per target.[13] A breakout of projected target costs based on a target factory economics model is shown in Figure 3.5.

Conclusion 3-4: Target fabrication at the quality and production rate needed appears possible with continued development.

---

[13] See, for example, D.T. Goodin, A. Nobile, J. Hoffer, et al., 2003, Addressing the issues of target fabrication and injection of inertial fusion energy, *Fusion Engineering and Design* 69: 803-806; R. Miles, et al., 2009, "LIFE Target Fabrication Costs," LLNL-TR-416932; and R. Miles, J. Biener, S. Kucheyev, et al., 2008, "LIFE Target Fabrication Research Plan," LLNL-TR-408722.

Scientific and Engineering Challenges and R&D Priorities

Target Fabrication

The scientific challenges to IFE target fabrication lie primarily in understanding the physics behind the specifications for inertial fusion target requirements: sphericity, uniformity and smoothness (How good is good enough?), and understanding the physics and chemistry behind the ability to achieve those requirements (Which physical processes control sphericity, uniformity, and smoothness?) Experiments with IFE targets at the NIF can help provide the physics understanding. The engineering challenges lie in selecting and developing materials that can achieve these requirements and in developing the processes and equipment needed to do so reliably and repeatedly with very high yield at reasonable cost.

The specific requirements appear at present to include these:

- The ability to fabricate IFE targets that meet specifications such as those for indirect drive:

  — Capsules with 4 mm diameter, <1 μm sphericity, ~100 μm wall with <0.5 μm Δw, <200 Å rms surface smoothness, and a surface power spectrum below the NIF capsule profile.
  — Hohlraums fabricated to ≤10 μm accuracy. Targets assembled to ≤10 μm accuracy.

  and those for direct drive:

  — Foam shell capsules with ~150 μm thick with <0.5 μm Δw and ~4 mm diameter with <1 μm sphericity. Foam density ≤100 mg/cm^3 with cell size <1 μm. A seal coat[14] on top of the capsule having a 1-5 μm wall with <0.5 μm Δw, <200 Å rms surface smoothness, and a surface power spectrum meeting the NIF/NIC required profile.

- A projected cost of mass-producing IFE targets for a power plant of ≤$0.50 each.

The objectives of IFE target fabrication R&D must be to understand the physics behind the specifications for inertial fusion target requirements and understand the physics behind the ability to achieve those requirements to such a depth that

---

[14] The seal coat surface for the direct drive capsule both seals the capsule and facilitates its injection into the target chamber without going out of specification by the time it reaches the center.

target materials can be selected and/or developed that meet target specifications, and processes and equipment can be developed to do so reliably and repeatably with very high yield at reasonable cost.

Target Injection at High Repetition Rates

After the targets have been fabricated they must be injected into the chamber. For laser drivers and accelerators, several methods of ballistic injection have been suggested, including gas guns and electromagnetic accelerators. For present pulsed-power fusion system designs, the targets are attached directly to the end of a transmission line. In this case, the targets and a replaceable transmission line are inserted into the chamber mechanically. Here, the committee considers only ballistic injection.

Gas guns have been built at LBNL and at General Atomics (Figure 3.6). These have been used to accelerate surrogate targets to high velocity (>100 m/s). In the case of direct drive, the targets must be carried by some kind of sabot to protect the target as it is accelerated in the gun barrel and injected into the chamber. The

FIGURE 3.6 Inertial fusion energy target gas-gun injection experiment. SOURCE: General Atomics.

sabot is removed either mechanically (with a spring) or magnetically. The gas gun experiments have demonstrated high-repetition-rate injection, including separation of the sabots from the targets, in a burst mode.[15] In these experiments, the placement accuracy at a distance of 20 m was about 10 mm. This 10 mm includes the contributions from the accuracy of the gun and from the separation of the target from the sabot. Estimates of the placement accuracy for indirectly driven targets (no sabots required) are much better than 10 mm. This is adequate for subsequent target tracking and beam steering, as discussed in the next section.

In summary, one can unquestionably build devices to inject the targets at adequate velocities and repetition rates. The remaining challenges are associated with wear and long-term reliability and durability, particularly in a fusion environment.

Conclusion 3-5: Target injection techniques have been developed in the laboratory that are adequate for subsequent target tracking and steering and that appear to be scalable to meet the inertial fusion energy requirements for speed and accuracy.

Target Tracking and Driver Pointing

The uncertainty in position of the targets when injected is much larger than the alignment precision of the driver beams relative to the target needed for ignition. Typically the required alignment precision is approximately 20 μm for both laser and ion direct drive.[16] For NIF-like, indirectly driven targets, the required precision is approximately 80 μm. For ion-beam indirect drive, the requirement is calculated to be 100 to 200 μm, depending on the size of the hohlraum. Given this situation, it is necessary to track the position of the target and to point the driver beams at the target. At least two methods of target tracking have been demonstrated. One tracks the shadow of the target using light-sensitive sensors. The other relies on the reflection ("glint") off the target. A scaled experiment performed by the University of California at San Diego and General Atomics demonstrated a beam alignment of 28 μm.[17] An alignment precision of 28 μm is nearly good enough, even for direct drive. Improvement to 20 μm seems possible, although shock-ignition targets may require still more precise alignment. The remaining challenge is to scale the technique to full size and full target velocity and demonstrate that it works reliably in a fusion environment. In a fusion environment one will undoubtedly have to

---

[15] D.T. Goodin, A. Nobile, J. Hoffer, et al., 2003, Addressing the issues of target fabrication and injection of inertial fusion energy, *Fusion Engineering and Design* 69: 803-806.

[16] L.C. Carlson, M.S. Tillack, J. Stromsoe, et al., 2010, Completing the viability demonstration of direct-drive IFE target engagement and assessing scalability to a full-scale power plan, *IEEE Transactions on Plasma Science* 38 (3): 300-305.

[17] Ibid.

deal with rapidly changing temperatures, mechanical vibration, and degradation of components by radiation.

The pointing of laser beams is usually done mechanically using a rapidly moving optical element. For accelerators, the beams can be pointed by pulsing relatively weak dipole magnets. For the beam parameters usually associated with ion indirect drive, this technique does not appear to be challenging. On the other hand, it may be necessary to put a significant energy spread on the ion beams to achieve the beam pulse durations needed for shock ignition or fast ignition. Energy spread produces dispersive effects in magnetic fields, so more work is needed to establish pointing feasibility for these options.

Conclusion 3-6: Target tracking and laser-beam-pointing methods that are adequate for indirect drive have been developed in the laboratory; direct drive will require higher precision.

Target Survival Under Hostile Conditions

The targets must survive injection into the target chamber and retain their precise dimensions, surface finish, and other characteristics until they are ignited by the driver beams. The insults they may sustain include acceleration in a gun, separation from a sabot, thermal radiation loads from the chamber walls, thermal and aerodynamic loads from residual gas in the chamber, and condensation of residual gas on the cryogenic target. The conditions are very challenging.

All high-gain target designs require cryogenic solid or liquid fuel and must remain at low temperature (<20 K) until they are fired. In contrast, the temperature of the chamber wall might be approximately 800 K, and the temperature of any gas in the chamber could be much higher. Indirectly driven fuel capsules are protected and insulated by the hohlraum. Numerical simulations indicate that these fuel capsules will survive even if there is significant gas in the chamber. Consequently, the LIFE power plant study, based on indirect drive, adopts gas wall protection. The chamber is designed to contain about 6 mg/cm^3 of Xe to protect the first wall and optical elements from photons and other target debris. Directly driven targets could not survive in such an environment, so the chambers chosen for these targets are usually designed to operate at chamber gas densities that are typically about three orders of magnitude lower. Under these lower-pressure conditions, calculations and some experiments indicate that the targets will survive at achievable injection velocities, even if the sabot carrying the target is stripped from the target as the target leaves the barrel of the injector and enters the chamber.[18] The implications

---

[18] J.D. Sethian, NRL, "Integrated Design of a Laser Fusion Target Chamber System," Presentation to committee on June 15, 2011.

for chamber design are discussed in the next section, entitled Chamber Technology. If it turns out to be highly desirable to have some kind of gas or liquid wall protection, it may be possible to delay the separation of the target and sabot until the target is very near the center of the chamber. In all cases, continued development of concepts and more experimental verification of target survivability in the expected chamber environment are needed.

Finally, the survivability issues for indirectly driven heavy-ion fusion and pulsed-power fusion appear to be less serious than the corresponding issues for laser fusion. Ion beams can penetrate the hohlraum wall so no laser entrance holes are required. For pulsed-power fusion, the target is usually part of a relatively massive transmission line that is placed into the chamber.

Conclusion 3-7: Analysis of target survival during injection into the target chamber indicates that survival of indirect-drive targets appears to be feasible. Further combined development of target and associated chamber systems will be needed to assure survival of direct-drive targets.

Recycling of Target Materials

All targets produce radioactive materials—unburned DT fuel if nothing else—that must be recycled. Nevertheless, targets for laser direct drive produce orders-of-magnitude less high-Z material than indirectly driven targets for both lasers and ion beams. Although the indirectly driven targets have the advantage in terms of injection, direct drive has the advantage in terms of recycling. Most direct-drive (actually mixed-drive) ion targets also contain significant quantities of higher-Z material. In the case of pulsed-power fusion, the target materials themselves are dwarfed by the transmission line structure that is destroyed on each pulse.

There is currently little agreement on how to handle the high-Z materials such as Pb, Au, and Pd. These materials will be activated to some extent and will have to be considered as radioactive waste. Some researchers believe that it is preferable to use new material, such as lead, for each target.[19] In this case, there is a significant waste stream but it is only mildly radioactive. In contrast, the LIFE team proposes to recycle the lead used for the hohlraums.[20] All surfaces in the reactor and vacuum chamber are designed to operate at temperatures exceeding the melting point of lead. The molten lead is collected and recycled. For liquid-wall chambers using

---

[19] L.A. El-Guebaly, P. Wilson, and D. Paige, 2006, Evolution of clearance standards and implications for radwaste management of fusion power plants, *Fusion Science and Technology* 49: 62-73.

[20] M. Dunne, E.I. Moses, P. Amendt, et al., 2011, Timely delivery of laser inertial fusion energy (LIFE), *Fusion Science and Technology* 60: 19-27; and J.F. Latkowski, R.P. Abbott, S. Aceves, et al., 2011, Chamber design for the laser inertial fusion energy (LIFE) engine, *Fusion Science and Technology* 60:54-60.

lithium or molten salt, the hohlraum materials would have to be removed from the liquid. There are a number of trade-offs involved in the choice of hohlraum material. Some materials are better than others in terms of target performance. Some are better in terms of activation, toxicity, and cost. Finally, some are easier to separate from the chamber liquid.

For IFE concepts with wetted or liquid wall chambers, it may be possible to make the targets from materials that are constituents of the chamber coolant. Lead hohlraums for use with LiPb coolants and frozen-salt hohlraums with a high-Z liner for use with liquid-salt coolants may be possible.

There has been significant research on nearly all of the issues associated with handling and recycling the target materials.[21] Determining the optimal methods and materials and demonstrating commercial feasibility remains an important challenge. Many of the topics associated with the recycling of tritium and other target materials will be discussed later in this chapter.

Conclusion 3-8: Target materials recycling issues depend strongly on the inertial fusion energy concept, the target design, and the chamber technology. Direct-drive targets have fewer concerns in the area of recycling and waste management; indirect-drive target materials handling, recycling, and waste management will need further development.

Path Forward

Each inertial fusion concept—direct-drive lasers, indirect-drive lasers, heavy ion beams, and pulsed power—will require its own specific target. Each of these will require target fabrication techniques for mass production. The targets for each IFE concept may have different materials and characteristics for injection, tracking and survival in the target chamber. While there may be some opportunities for synergy between different target technologies, the following R&D steps will be required for each inertial fusion concept.

Near Term (<5 Years)

- Work with target designers to jointly agree on designs that promise high gain, practical fabrication, good mechanical strength, and good thermal robustness.
- Continue development, begun under the ICF program, of physics models of the formation of small hollow spheres, subsequent DT layering, and other fabrication processes.

---

[21] L.A. El-Guebaly, P. Wilson, and D. Paige, 2006, Evolution of clearance standards and implications for radwaste management of fusion power plants, *Fusion Science and Technology* 49: 62-73

- Demonstrate gain using prototype targets made of commercial IFE materials with expected fabrication specifications and tolerances on the NIF.
- Quantify detailed target requirements and manufacturing tolerances.
- Select and demonstrate target fabrication techniques for low-cost mass production.
- Develop characterization and statistical sampling techniques needed for IFE mass production.
- Demonstrate DT filling and layering/wicking protocols suitable for IFE targets.
- Develop an IFE target factory conceptual design and cost estimate. Conceptualize a target factory test facility with single units of small machines, leading to a target factory with multiple units of larger machines of similar design.
- Continue laboratory-scale development of target injection and tracking techniques, including studies of target survival during injection and transport into a simulated target chamber.
- Investigate target materials recycle and waste management issues.

Medium Term (~5-15 Years)

- Test IFE target concepts in the NIF; determine sensitivity to target fabrication parameters and tolerances.
- Design a target factory and injection and tracking system to supply targets to the first IFE demonstration facility.
- Put in place target material recycling and/or waste stream management processes.

Long Term (>15 Years)

- Develop the technologies for construction of a commercial target factory for an IFE power plant.
- Update techniques and factories for the mass fabrication of targets to reflect the latest target designs.

Conclusion 3-9: An inertial fusion energy program would require an expanded effort on target fabrication, injection, tracking, survivability, and recycling. Target technologies developed in the laboratory would need to be demonstrated on industrial mass production equipment. A target technology program would be required for all promising inertial fusion energy options, consistent with budgetary constraints.

CHAMBER TECHNOLOGY

Background and Status

An IFE system will require the means to extract and utilize the energy produced by the fusion events that take place inside the reaction chamber; the ability to breed, extract, and process the tritium fuel; and the ability to maintain these systems in a timely manner. The systems must allow for delivery of the driver energy to the target and must ensure that the chamber can withstand the target emissions over timescales of a year or more. All this must be done in a way that meets the safety and environmental goals for a commercial energy system.

This section discusses the issues, challenges, and R&D needed for chamber options for IFE while other sections in this chapter discuss the related issues of materials, tritium systems, and safety and environmental topics.

A number of IFE design studies have been carried out that, while preliminary, shed light on the key features of the chambers of IFE systems. These include the OSIRIS/SOMBRERO[22] and Prometheus[23] studies that developed reactor designs for laser and heavy-ion drivers. There are also other studies on heavy-ion chambers from HIBALL,[24] HYLIFE,[25] and the Robust Point Design and Hylife-II studies,[26] while information on pulsed-power reactors has also been reviewed.[27] The most recent design efforts are the HAPL direct drive laser design[28] and the LIFE indirect-drive laser design.[29] The information that follows in this section is a composite of the information in these references.

---

[22] DOE, 1992, *OSIRIS and SOMBRERO Inertial Fusion Power Plant Designs*, DOE/ER-54100-1.

[23] DOE, 1992, *Inertial Fusion Energy Reactor Design Studies Prometheus-L and Prometheus-H*, DOE/ER-54101.

[24] B. Badger, K. Beckert, R. Bock, et al., 1981, *HIBALL—A Conceptual Heavy Ion Beam Fusion Reactor Study*, UWFDM-450, University of Wisconsin at Madison, and KFK-3202, Kernforschungszentrum Karlsruhe.

[25] J.A. Blink, W.J. Hogan, J. Hovingh, W.R. Meier, and J.H. Pitts, 1985, *The High Yield Lithium Injection Fusion Energy (HYLIFE) Reactor*, UCRL-53559, LLNL.

[26] S.S. Yu, W.R. Meier, R.P. Abbott, et al., 2003, An updated point design for heavy ion fusion, *Fusion Science and Technology* 44(2): 266-273.

[27] See C.L. Olson, 2005, "Z-Pinch Inertial Fusion Energy," *Landolt-Boernstein Handbook on Energy Technologies*, VIII/3: 495-526, Springer-Verlag, Berlin; and G.E. Rochau and C.W. Morrow, 2004, *A Concept for a Z-Pinch Driven Fusion Power Plant*, SAND2004-1180.

[28] J.D. Sethian, D.G. Colombant, J.L. Giuliani, et al., 2010, The science and technologies for fusion energy with lasers and direct-drive targets, *IEEE Transactions on Plasma Science* 38(4): 690-703.

[29] M. Dunne, E.I. Moses, P. Amendt, et al., 2011, Timely delivery of laser inertial fusion energy (LIFE), *Fusion Science and Technology* 60: 19-27; and J.F. Latkowski, R.P. Abbott, S. Aceves, et al., 2011, Chamber design for the laser inertial fusion energy (LIFE) engine, *Fusion Science and Technology* 60:54-60.

The technology for the reactor chambers, including heat exhaust and management of tritium, involves difficult and complicated issues with multiple, frequently competing goals and requirements. Understanding the issues and the options for resolution is important for establishing that credible pathways exist for the commercialization of IFE, and this will require significant effort. Understanding the performance at the level of subsystems such as a breeding blanket and tritium management, and integrating these complex subsystems into a robust and self-consistent design will be very challenging.

The two main classes of reaction chamber are those with solid walls and those with liquid walls. The key feature of liquid wall chambers is the use of a renewable liquid layer to protect chamber structures from target emissions. Two primary options have been proposed and studied: wetted-wall chambers and thick liquid-wall chambers.

With wetted-wall designs, a thin layer of liquid on the inside of the wall shields the structural first wall from most short-range target emissions (X-rays, ions, and debris) but not neutrons. Various schemes have been proposed to establish and renew the liquid layer between shots, including flow-guiding porous fabrics, porous rigid structures and thin film flows. Similarly, various schemes have been proposed to protect beam ports and final optics. The thin liquid layer can be the tritium-breeding material (e.g., FLiBe, PbLi, or Li) or another liquid such as molten Pb. Such thin layers will contribute to tritium breeding, but not significantly.

With thick-liquid-wall designs, liquid jets are injected by stationary or oscillating nozzles to form a neutronically thick layer (typically with an effective thickness of ~50 cm) of liquid between the target and first structural wall. Gaps are provided between the thick liquid flows for access by the driver beams. This is much easier to accomplish for indirect drive, which can have a biaxial or even uniaxial beam geometry, than for direct drive, which requires many driver beams to achieve drive symmetry. In addition to absorbing short-range emissions, the thick liquid layer degrades the neutron flux and energy reaching the solid material first wall, so that the structural walls may survive for the life of the plant (~30-60 yr). The thick liquid serves as the primary coolant and tritium breeding material. In essence, the thick liquid wall places the fusion blanket inside the first wall instead of behind the first wall. A significant potential advantage of thick liquid wall designs is that the neutron damage to chamber structures can be reduced considerably due to the shielding provided by the liquid. This allows for a reduction of the waste stream as the need for replacement of the chamber structures can be minimized, resulting in a simplification of the waste management requirements and improving availability. An example is shown in Figure 3.7, where the target and driver beams enter the chamber biaxially between thick liquid flows. It is also possible, in principle, to have centrifugally maintained thick liquid walls.

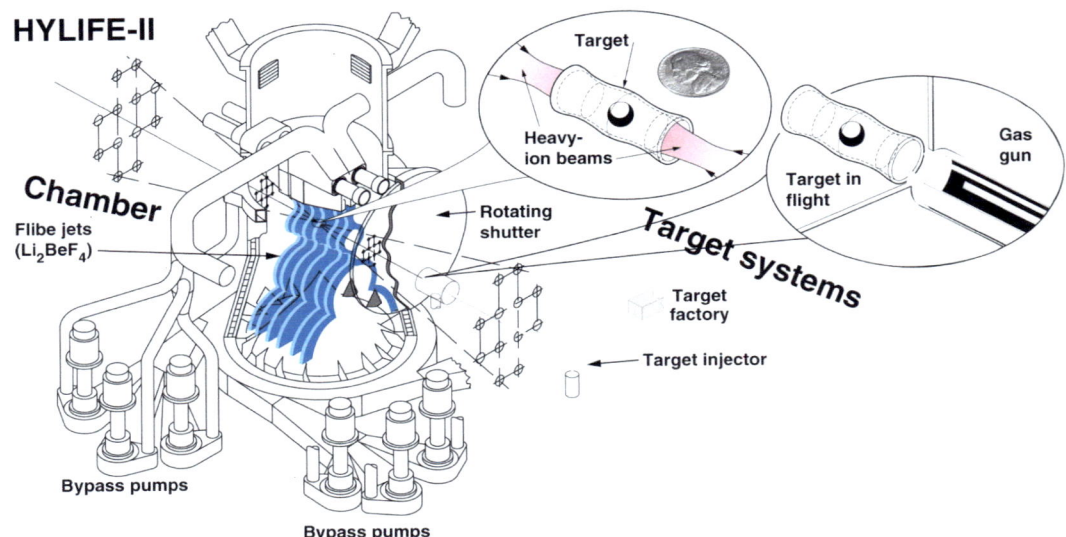

FIGURE 3.7 Thick-liquid-wall chamber for heavy ion fusion. SOURCE: LBNL.

Solid- or dry-wall chambers are expected to be compatible with laser-beam or ion-beam entrance into the chamber. If the dry wall chamber is evacuated or has a gas fill of no more than a few tens of millitorrs (at room temperature), then it may be possible to have easier target injection, target tracking, target survival, high-fidelity laser propagation, restoration of chamber conditions for the next shot, and gas reprocessing (e.g., cooling and target debris removal).

Dry-wall chambers, which have no constraints for liquid film or liquid jet geometry, should be able to accommodate the illumination geometry for either direct-drive or indirect-drive targets. For laser drivers, chamber designs have been proposed to deal with target emission from either direct-drive (e.g., HAPL[30]) or indirect-drive (e.g., LIFE[31]) targets. An example is shown on Figure 3.8.

Wetted-wall chambers could be compatible with either direct-drive or indirect-drive illumination, but there are some advantages to indirect drive since it would be possible to configure the beam paths from the sides and this could reduce the chance of liquid reaching the final optics. The thin liquid layer would be able to withstand short-range ion, X-ray, and debris emissions from either direct-drive or indirect-drive targets.

---

[30] J.D. Sethian, D.G. Colombant, J.L. Giuliani, et al., 2010, The science and technologies for fusion energy with lasers and direct-drive targets, *IEEE Transactions on Plasma Science* 38(4): 690-703.

[31] M. Dunne, E.I. Moses, P. Amendt, et al., 2011, Timely delivery of laser inertial fusion energy (LIFE), *Fusion Science and Technology* 60: 19-27.

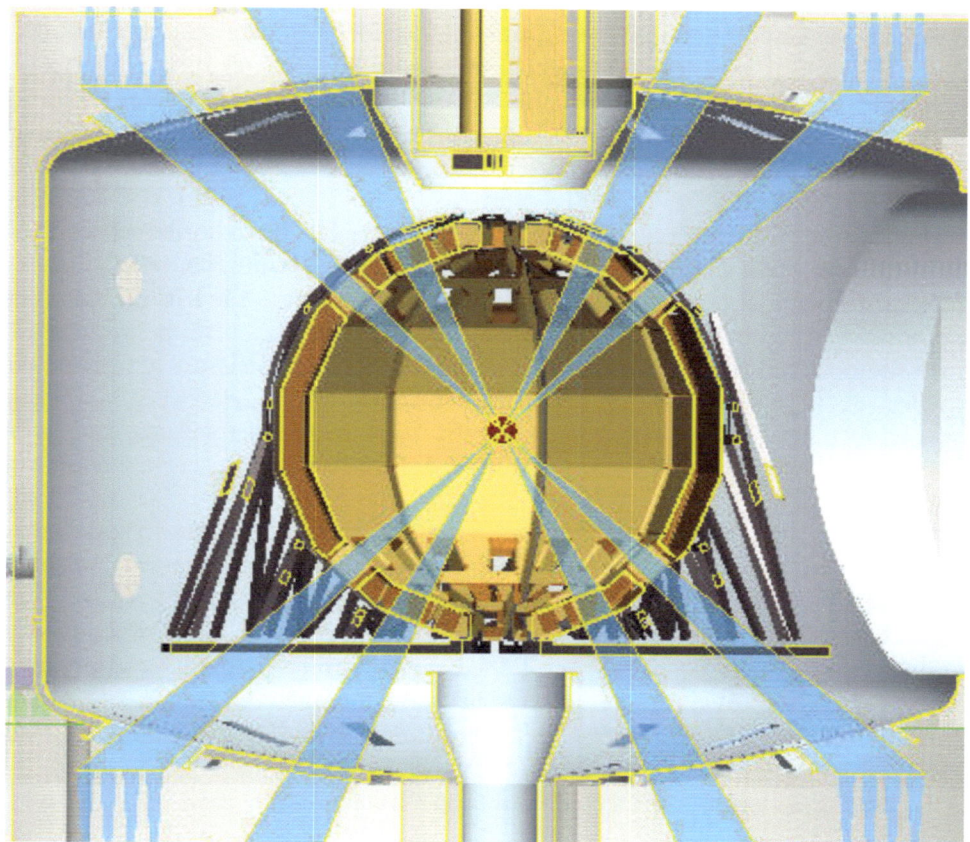

FIGURE 3.8 Example of a dry-wall chamber concept developed for the LIFE project. SOURCE: M. Dunne, E.I. Moses, P. Amendt, et al., 2011, Timely delivery of laser inertial fusion energy (LIFE), *Fusion Science and Technology* 60: 19-27.

There are additional issues associated with the incorporation of liquids into the reaction chamber. Thick liquid walls are likely only compatible with indirect-drive targets unless extraordinary measures are taken to provide a thick shielding region between up to hundreds of beam paths. The thick liquid layer should withstand the energy pulse of the target emissions. Indirect drive and magnetically driven direct drive with thick liquid wall chambers would be the primary choices at present for heavy-ion and pulsed-power drivers, respectively.

It is important to note that the pulse repetition rates very much affect the chamber issues. Such rates vary from 16 Hz for some laser drivers, to around 5 Hz for heavy ion driver concepts, and to about 0.1 Hz for pulsed power concepts. For example, increased repetition rates imply higher target injection speeds that can increase the heat load to the cryogenic targets in gas-filled chambers. Increased repetition rates will also mean less time to clear the chamber for the next shot and may nessecitate larger pumping ports. Higher rates also reduce the time available for cooling of the chamber gas between shots.

All fusion concepts, both IFE and MFE, must provide for tritium self-sufficiency in order to have a closed fuel cycle needed for commercial success or even large-scale test facilities. This covers a range of issues, including performance of the target (especially the tritium burnup fraction), the tritium breeding potential of the blanket, tritium recovery and storage, and tritium inventories, including tritium hold-up in the walls of the chamber. These issues are discussed in more detail in the following section on tritium production, recovery, and management. In general, IFE will greatly benefit from the long experience and large investments being made in the worldwide MFE program on tritium breeding and handling.

IFE has a potentially advantageous feature in that the driver system and chamber system are not necessarily closely connected together. Furthermore, it appears to be possible to take advantage of the modular nature of at least some of the driver candidates. These features offer potential benefits in terms of plant maintenance and availability. Further, this decoupling and ability to test modular components without building the entire reactor system should reduce the cost and the time needed to qualify IFE components. For the chamber, periodic replacement or repair would be undertaken—hopefully, only every few years.

These considerations lead to the following conclusion:

Conclusion 3-10: The chamber and blanket are critical elements of an inertial fusion energy power plant, providing the means to convert the energy released in fusion reactions into useful applications as well as the means to breed the tritium fuel. The choice and design of chamber technologies are strongly coupled to the choice and design of driver and target technologies. A coordinated development program is needed.

Scientific and Engineering Challenges and Future R&D Priorities

There are, in general, significant threats to IFE chambers, particularly for those concepts that utilize solid walls. These threats include surface blistering and exfoliation due to ion implantation, near-surface ion and thermal damage, dust creation and material redeposition, cyclic thermomechanical stresses, volumetric fusion neutron and gamma-ray damage, and nuclear heating. Some of these issues are similar to those faced by MFE concepts, although the inherent pulsed nature of IFE poses unique challenges. Of special concern to IFE laser concepts is the damage to laser system final optics. These issues are discussed in more detail in the next section, Path Forward.

The key challenge for a dry-wall concept is to establish a configuration that can repeatedly withstand the typically 300 million high-energy pulses per year of X-rays, ions, and neutrons coming from the target. This threat spectrum depends on the target design. For almost all IFE targets, roughly 70 percent of the fusion energy is released as neutrons. For a direct-drive target, typically 28 percent comes out in ions and 2 percent in X-rays. For an indirect-drive target, the non-neutron ratio is roughly inverted: 25 percent comes out in X-rays and 5 percent in ions.

The basic requirements for the chamber to operate at the necessary pulse repetition rates (which can vary from ~10 Hz to 0.1 Hz) are, after each shot:

- Reestablish chamber conditions that allow for the delivery of the target with the required precision and without damaging the integrity of the target.
- Reestablish chamber conditions that allow for delivery of the driver energy to the target including high-repetition-rate target tracking and beam pointing for lasers and heavy ion drivers.
- Reestablish in-chamber conditions that may be used to protect chamber structures from target emissions (e.g., liquid films, liquid jets, and gases) and/or assure survival of the first wall subjected to pulsed energy loads.

For dry-wall chambers, an important issue is target heating during injection due to thermal radiation from the hot chamber wall. There may also be some residual target materials and potential gas propellant from previous shots in the chamber that could add to target heating and affect its trajectory. The use of infrared reflective coatings and/or protective sabots on the target may reduce heating by the wall. For gas-filled chambers, the gas fill dominates in-chamber conditions and will have a greater impact on target heating and trajectory than the walls of evacuated chambers. It will be necessary to limit the gas density and chamber radius to values that allow the target to survive.

For liquid-wall chambers, the liquid vapor filling the chamber contributes to target heating and impacts the trajectory. Liquid drops, if present, must not interfere with target delivery. The protective liquid layers and jets must be reconstituted after the disruptive effects of the target emissions. For pulsed-power concepts, the key issue is the mechanics of delivering the combined recycled transmission line and target system. It will be necessary to reset the liquid sheets to allow subsequent target injection in 1-10 s.

For direct-drive targets (laser or heavy-ion concepts), uniform beam delivery could also be affected by residual vapors, droplet formation, and turbulence from remnant target materials. For laser drivers, the final optics are in direct line of sight of target emissions and thus subject to possible degradation from target debris, thin-film deposition, and neutron, X-ray, and charged-particle damage. It may be possible to use magnetic deflection of ions to protect the entrance ports and final optics. For gas-filled chambers, the buffer gas may protect the final optics from short-range target emissions. In any event, it will be necessary to choose final optics that are least susceptible to surface perturbation and alignment error.

The first wall is subject to threats from the X-rays and ions. With no gas in the chamber, the X-rays are delivered in very short (a few nanoseconds) pulses. Their energies range from 0.1 to 100 keV, so their penetration depth is 10 to 200 μm, depending on the wall material. The X-rays from direct drive are harder, more penetrating, and less numerous than those expected from indirect drive, so the instantaneous wall temperature rise is lower. The ions, because of their slower velocity, reach the wall several microseconds after the X-rays. In addition, their energy is imparted to the wall on a timescale of a few microseconds, owing to the different energies and species of the ions. The ion spectrum depends on the type of target but will always have the hydrogen isotopes, helium, and carbon as well as the hohlraum species with indirect drive. Generally, the ions deposit their energy and implant within a few microns of the surface, giving a temperature spike and potentially causing first wall material erosion.

Lead is a prime candidate for and example of a particular hohlraum material. It has been selected as both the high-Z and substrate material for indirect-drive targets. Lead has a high opacity to thermal X-rays (thus giving good driver coupling efficiency), is inexpensive and widely available, is compatible with laser beam propagation, and has a favorable melting point and vapor pressure curve that support removal from the chamber. In the LIFE design example, each target contains approximately 3 g of lead, which amounts to a daily throughput of about 4 tonnes. This material would be collected and recycled into future targets. The target chamber xenon fill gas remains sufficiently hot between shots such that the vast majority of lead will remain in the vapor phase. Some of the lead will reach the first wall and blanket structures, where it can condense. Condensed lead will either run down the wall to the debris collection/gas exhaust port at the bottom of the chamber, or it will drip. Gas pumping occurs at the bottom of the fusion chamber. This gas is processed to remove lead, hydrogen isotopes, etc., and is then recompressed for injection into the low-pressure vacuum chamber. Gas injection occurs near the final optics over a relatively small area so that an increased gas velocity is achieved. This gas flow inhibits the flow of particles or droplets to the final optic.

There are more avenues to alleviate the effects of ions than the effects of X-rays, because ions are slower, deposit energy over a longer time, and have an electrical charge that allows them to be diverted. For an indirect drive target, with the much higher fraction of X-rays in the threat spectrum (25 percent vs. 2 percent in direct-drive systems), the volumetric X-ray power deposition is sufficient to melt and possibly even vaporize the chamber wall surface. The timescale for the deposition energy from these X-rays is much shorter than the energy transport timescale in materials so that all the energy is absorbed in the materials' surface layers, which leads to repetitive melting and ablation. For example, the surface of a tungsten wall at 10 m radius would be heated to over 6000°C, well past the tungsten melting point, with an indirect-drive target that releases 200 MJ/shot. Thus, any indirect-drive target requires some type of replenishable buffer to protect the solid wall. Options include thin liquids, thick liquids, or a buffer gas. For a direct-drive target, the energy in the X-rays is relatively small, so the X-rays from a 200 MJ target heat up a 10-m-radius tungsten wall to only 1000°C. The ions, when they arrive later over a longer pulse, heat the wall to 1650°C. This is below the melting point of tungsten but still it pushes past the recrystalization temperature and may lead to the formation of cracks.

The dry-wall concepts must also account for the time-averaged power density that requires that the target-facing materials be actively cooled, resulting in thermal stresses in the first wall structure. This may limit the thickness of the chamber facing materials because the surface temperature needs to be lowered before the next pulse to avoid thermal limits at the surface.

Material options for the first wall of solid wall concepts include graphite or SiC composites, as well as refractory metals such as tungsten. Various concepts for engineered materials have been proposed, such as carbon brush structures, tungsten foam, vacuum-sprayed nanoporous tungsten structures, and diffusion-bonded or plasma-sprayed tungsten on ferritic steels.

The use of liquid walls alleviates many of these solid wall concerns but introduces other issues, such as the need to manage vaporization of the liquid and subsequent clearing in the chamber, uniform liquid wetting and refilling at 5-10 Hz, liquid mobility, and the effect of splashing on optics.

Despite the many competing requirements and complicated interactions of the technologies needed for IFE chambers, plausible solutions and self-consistent designs have been put forward for all IFE concepts in the design studies that have been done. Table 3.1 provides a summary and review of the chamber concepts and main issues.

Conclusion 3-11: Chamber and blanket technologies involve a broad range of very challenging and complex interrelated issues rooted in many science and engineering disciplines. Resolving these issues will take a dedicated effort over many years of research and development.

From the scientific and engineering challenges identified in the previous subsection, one can develop a set of demanding R&D objectives that must be addressed for realizing the potential of IFE as an energy system. In general, work on these issues is not being funded at present.

TABLE 3.1 Summary of Inertial Fusion Energy Chamber Concepts and Issues

	Thick Liquid Wall	Solid Wall	
		Protective Gas	Vacuum
	Heavy Ions (HI) Pulsed Power (Z)	Laser Indirect Drive	Laser Direct Drive
Primary advantage	Fewer materials issues with X-rays, ions, or neutrons. Thick liquid also breeder/coolant.	Fewer first wall X-ray or ion material issues.	Simplicity.
Primary challenge	Chamber clearing, target placement.	Chamber clearing, laser propagation.	First wall resistance to helium retention, surface morphology change, and mass loss.
Target survival	Hohlraum thermal insulation.	Hohlraum thermal insulation.	IR protective layer, start target cold.
Driver/target coupling	(HI) Accurate target injection. (Z) Target part of RTL: automatically aligned.	Inject target close enough to chamber center to allow laser mirrors to be steered to required accuracy.	Inject target within 1 cm of chamber center, detect glint from target, and steer laser mirror to required accuracy.
Resistance to emissions of X-rays, ions, and neutrons	Thick liquid resistant to all emissions, including neutrons.	6 μg/cm^3 xenon gas (760 mTorr at STP). Modeling: gas stops X-rays, reemits later peak wall T < 850°C.	Engineered tungsten or magnetic intervention.
Chamber recovery: rep-rate and clearing	(HI) Oscillating liquid jets sweep chamber (Z) Metal "waterfalls" protect walls; RTL obviates clearing.	Recycle 0.5% of gas between shots.	Evacuate the chamber; well within commercial technology.
Breeder/coolant	Thick liquid.	Lithium, behind first wall.	FLiBe or PbLi behind first wall.
Chamber repetition rate and clearing issues	(HI) Do oscillating jets sweep out enough ionized/atomized liquid for driver propagation and target injection? (Z) Demonstrate RTL concept with scaled experiments.	Target survival and adequate quality laser propagation through residual hot Xe or Xe/Pb gas/plasma.	Only gas load is from vaporized direct-drive target ~0.025 mTorr per shot.
Chamber chemistry issues	Proposed liquid: FLiBe also maybe Na. All are very reactive. Must stay "chemically locked up" when subject to X-rays, ions, and heat.	Effect of lead liquid / vapor (from hohlraum) on wall and optics. Deposition of carbon-tritium on "colder" surfaces.	Should be no chemistry issues with tungsten wall. Deposition of carbon-tritium on colder surfaces.
Other critical issues	(Z) RTL "insertion hole" needs protection from emissions	Target survival/laser focusing experiments	He retention; finish target warm-up

NOTE: RTL, recyclable transmission lines. SOURCE: J.D. Sethian, Communication to the committee on August 19, 2011.

Conclusion 3-12: At present there is no specific program in the United States addressing IFE chamber issues.

In general these R&D objectives, which may be one of the most important pacing items in the commercialization of fusion, include handling of the heat exhaust and waste heat for the driver, chamber, and balance-of-plant systems; development of radiation-resistant and affordable materials; development of tritium handling systems; hydrodynamics of thick liquid walls and response to fusion blast; management of repetitive shocks and fatigue effects for dry and wet walls; resolution of first-wall issues of erosion, helium blistering, tritium retention, and neutron damage; development of approaches for nuclear waste management and minimization approaches; resolution of IFE safety-related issues; and development of designs for durable chambers that resist damage from the repetitive pulsed emissions from the target.

Given that direct-drive targets may not tolerate sufficient gas to stop all of the emitted burn ions, direct-drive chambers must be designed to handle both the thermal pulse resulting from X-ray irradiation and ion implantation as well as erosion damage due to the ion flux itself. Alternatively, ions might be diverted magnetically.

The thick liquid wall chamber concepts may not require testing in high-neutron-fluence materials facilities. Instead, these types of chambers could be developed and tested using a combination of multiscale modeling, validation experiments, accelerated damage testing, and in situ monitoring, thus reducing the development time and cost of a IFE program.

Path Forward

Specific R&D for Liquid Walls

The key goals of R&D in this area would be to demonstrate the ability to create the protective liquid configuration and to determine the response of the liquid to the fusion yield, including response to neutron energy deposition. Specific tasks include the ability to mitigate shock and debris and to show that the protection can be reestablished prior to the next shot while assuring target and driver energy-delivery and the feasibility of cleaning and circulating the liquid at a sufficient time-averaged rate. Because the ablation and neutron heating occur on a timescale that is much shorter than hydrodynamic response, subscale tests with simulant fluids and nonfusion impulse loads could be used to test key issues of response and reestablishment of the liquid protection. The R&D goals for three time horizons follow.

Near Term (<5 Years)

Needed R&D activities include systems studies; liquid-jet hydraulics; wetted-wall hydraulics; ablation/venting/condensation; laser final optics protection; FLiBe and liquid metal chemistry, corrosion, and tritium recovery; and modeling and experiments to demonstrate repetitive target injection in simulated liquid-wall-chamber conditions.

Medium Term (5-15 Years)

Success would be experimental validation of models required to extrapolate to prototypical chamber conditions, coupled with integrated system designs meeting clearing rates and other metrics. Candidate thick liquid wall concepts in flow loops, including tritium extraction, would be tested. Presuming that thick-liquid-wall concepts will be found viable, during this period experimental activities would occur to provide engineering-design capability, including integrated ablation/venting/condensation experiments; integrated liquid hydraulics testing; and beam propagation experiments to study the effects of background gas density and residual liquid droplets on heavy-ion/laser beam propagation under prototypical chamber conditions.

Long Term (>15 Years)

The objective would be to develop liquid-wall target chambers operating at 0.1 to 10 Hz, to be made available for an IFE fusion test facility (FTF) and subsequent IFE demonstration and commercial fusion power plants.

Specific R&D for Dry Walls

Dry-wall concepts must be shown to allow propagation of both the cryogenic target and driver beams to the target chamber center; possess adequate component lifetime in the face of neutron and ion damage to chamber materials; and enable ease of maintenance to maximize high plant availability.

Near Term (< 5 Years)

Designs will be developed and tested for an integrated chamber and target injection system. The fundamental response of various candidate materials to a prototypical plasma (flux, energy spectrum, species spectrum) would be investigated, as well as the retention of tritium in these materials. Measurements of gas cooling and laser beam propagation through representative chamber gas mixtures would be carried out.

Medium Term (5-15 Years)

During this time a design of an IFE engineering test reactor with a dry-wall concept using available structural materials for the chamber would be carried out. Wall damage mitigation strategies would be evaluated, including these:

- Magnetic deflection of implosion ions;
- Buffering gas options (e.g., trade-offs between turbulence effects on target delivery and reducing the range of implosion ions); and
- Replenishment of wall surfaces (e.g., thin liquid surface coatings on capillaries).

Sufficiently rapid chamber clearing and protection of final optics would be demonstrated.

Long Term (>15 Years)

The overall objective would be to operate an FTF utilizing chamber materials that were qualified during the medium-term phase. Demonstration of chamber maintenance and long-term plant availability to commercial levels would be a key objective.

Related R&D

Components in the vicinity of any fusion chamber will be activated within a short time of the start of operation of the plant, so remote maintenance capability will be required. This requirement is not unique to IFE; rather, it is similar to that of MFE and fission reactors. The degree of remote maintenance will vary with chamber concept. For example, if the thick liquid wall chamber can last for the life of the plant, remote maintenance will not be required for that component. It may be prudent, however, to include full remote maintenance capability even if the particular design is expected to have minimal remote maintenance needs. Systems developed for MFE, including ITER, will benefit IFE in general.

While the configurations and constraints may differ significantly from MFE to IFE, there are many common issues and interests, such as performance of materials in a fusion environment; tritium breeding blankets; tritium concerns including recovery, processing, accountability, and minimizing inventory; operation at high temperatures; corrosion of materials in contact with liquid metals or molten salts; erosion and formation of particulates (dust); advanced computational tools for neutronics; remote maintenance; and radiation-hardened diagnostics and instrumentation for in-vessel components. Thus IFE should benefit greatly from the MFE program in these areas in both the United States and worldwide. Conversely, IFE research could also benefit MFE development.

These considerations then lead to two recommendations for IFE chamber technologies:

> **Recommendation 3-3: The development of a strategy and roadmap for a U.S. IFE program should include the needs of chamber and blanket science and technology at an early date. A significant investment in upgraded and new test facilities and supporting R&D will be required.**
>
> **Recommendation 3-4: The U.S. IFE chamber R&D program should closely monitor R&D progress in the national and international MFE programs and should look for opportunities for collaboration with these programs.**

MATERIALS

Background and Status

Although achieving controlled thermonuclear fusion at breakeven efficiency remains a challenge, there is a reasonable expectation that it will be attained eventually and so the committee will turn its attention to exploiting thermonuclear fusion as an energy source. To accomplish this it expects to encounter formidable materials-related problems that will likely require research to solve. Elsewhere in this report the committee discusses materials issues arising in the lasers, particle accelerators, and pulsed power systems that serve as drivers for the implosion of a deuterium-tritium (DT) target. Here it concentrates on the materials that are needed for capturing that explosive neutron, ion, and X-ray energy to make power and breed more tritium fuel. Other reaction chamber technology issues are discussed in the preceding section.

Following the target's implosion, 70 percent of the energy appears as high-energy (millions of eV) neutrons, mainly from the D + T reaction (14 MeV) but some at lower energies from the T + T and D + D reactions. The remainder of the energy is in the form of energetic ions and X-rays. For the direct-drive configuration, 28 percent of the energy is in the MeV ions that come from the alpha particles (helium), protons, tritons, and ^{3}He ions that accompany the neutrons in the nuclear reactions just listed. In addition, there are many lower-energy ions (carbon and metal ions) from the destruction of the target and the unburned DT fuel. The remainder of the energy from a direct-drive target (2 percent) is in the form of X-rays emitted by the target plasma heated by the charged fusion reaction products. In an indirect-drive implosion, these numbers are reversed—5 percent in ions and 25 percent in X-rays from the target and hohlraum.

To make useful power and future tritium fuel, we must capture and dissipate the energy of the neutrons, ions, and X-rays while simultaneously slowing the neutrons to thermal energies in order to breed tritium through the n + ^{6}Li nuclear reaction. Tritium is also produced by higher energy neutrons on ^{7}Li and ^{9}Be. This is where the challenges in material selection arise. Both neutrons and ions can damage the chamber materials, and this must be protected against or tolerated. Damage to the final stage of the laser optical elements, which have to have a line-of-sight visibility to the target, must also be minimized or nearly eliminated. For heavy-ion drivers, the accelerated ions can be deflected by magnetic fields, keeping the final beam focusing elements away from line of sight of the target, in principle shielding them from exposure to the neutrons, ions, and X-rays.

Scientific and Engineering Challenges and Future R&D Priorities

As noted earlier, in the indirect-drive configuration, the X-ray flash from the implosion will raise the wall temperature to a high level for a brief time (~6000°C for a 10 m chamber and 200 MJ release), enough to vaporize all solid or liquid wall materials. Obviously, such thermal cycling may lead to accumulated damage in the exposed materials. For this reason, a low-pressure, inert buffer gas such as helium can be used to fill the target chamber to reduce the thermal load on the wall. For a laser-based, direct-drive configuration, no appreciable buffer gas can be employed, but since the X-ray flux is lower, the metallic wall temperature rises only to about 1000°C. In this situation, however, in the absence of a magnetic field, the wall would be exposed to the full ion flux, which causes erosion by sputtering, and the implanted ions lead to near-surface (microns) damage (blistering etc.) and subsequent exfoliation of wall material. This produces an evolution of wall topography that may frustrate the use of nanostructured surfaces of materials such as tungsten or silicon carbide (SiC).

In addition, the repetitive thermal cycling of the materials (for example, below and above the recrystallization temperature) can seriously degrade the viability of the material even if the temperature increase is below that which causes fundamental phase transitions. Liquid surfaces present the possibility for self-healing; however, even liquid walls are subject to sputtering, evaporation, small-particle ejection, and aerosol formation. By putting magnetic coils outside the target chamber, the resultant magnetic field can be used to prevent ions from reaching the wall and divert them into shielded regions, which is another way of reducing damage to a large portion of the target-facing wall. A decade ago, a comprehensive report was written on the materials issues associated with IFE[32] and has been made available

---

[32] L. Snead, N.M. Ghoniem, and J.D. Sethian, 2001, "Integrated Path for Materials R&D in Laser Inertial Fusion Energy (IFE)," Internal memorandum, Naval Research Laboratory, August.

to the committee. Because it has abstracted from that source some of its comments on dry-wall chambers and final optical elements, the reader is encouraged to look there for more details.

Some damage to wall and optical elements will be similar to damage expected in magnetic confinement fusion as far as total neutron radiation fluence is concerned; however, it is well known that there are significant dose-rate effects that will be associated with the pulsed nature of inertial fusion. Data on these effects are sparse, and a continued R&D program on IFE must necessarily include provision for the facilities and experiments needed to probe this extreme radiation environment, especially the 14 MeV neutrons. If dedicated facilities are not provided for these studies, then it is likely that the first prototypes of IFE plants will be needed to perform the final experiments of the materials selection program.

Most of the existing studies have focused on the damage-rate effects associated with accelerated damage studies using ion- or electron-irradiation sources compared to fission reactor sources (both in steady state). There are no fusion neutron sources with sufficient neutron flux to do high-fluence neutron irradiation testing. Testing can be done using ions or with fission neutrons. Modeling[33] and experimental studies[34] have specifically examined the effects of IFE-relevant pulsed and steady-state irradiation conditions. These studies indicate that microstructural differences between pulsed and steady-state may occur, but some investigators think these differences are relatively small compared to other experimental variables such as damage level and irradiation temperature (for example, a change in temperature of 50°C typically has a bigger effect than the difference between pulsed and steady-state irradiation). There is not general agreement on this issue, so such effects need to be investigated in detail.

Another critical issue is the ability of the target-facing materials to capture and retain unburned tritium fuel. For safety reasons—for example, no site boundary evacuation—the present ITER design considerations indicate that <1 kg of tritium fuel will be allowed to be retained in the target-facing material.[35] A 2.5-GW thermal DT fusion plant burns about 0.5 kg/day of tritium, with the expected burn fraction

---

[33] N.M. Ghoniem and G.L. Kulcinski, 1982, A critical assessment of the effects of pulsed irradiation on the microstructure, swelling, and creep of materials, *Nuclear Technology-Fusion* 2: 165-198; H. Trinkaus and H. Ullmair, 2001, Does pulsing in spallation neutron sources affect radiation damage?, *Journal of Nuclear Materials* 296: 101-111; R.E. Stoller, "The Effect of Point Defect Transients in Low Temperature Irradiation Experiments," Presentation at ICFRM10, Baden-Baden, October 2001.

[34] E.H. Lee, N.H. Packan, and L.K. Mansur, 1983, Effects of pulsed dual-ion irradiation on phase transitions and microstructure in Ti-modified austenitic alloy, *Journal of Nuclear Materials* 117: 123-133; J.L. Brimhall, E.P. Simonen, and L.A. Charlot, 1983, Void growth in pulsed irradiation environment, *Journal of Nuclear Materials* 117: 118-122.

[35] B. Lipschultz, X. Bonnin, G. Counsell, et al., 2007, Plasma-surface interaction, scrape-off layer and divertor physics: Implications for ITER, *Nuclear Fusion* 47: 1189-1205.

of 30 percent. Therefore, 1 kg of tritium fuel is incident on the target-facing materials every day of operation. To assure that the IFE plant continues to operate for more than 1 year, the materials cannot retain more than ~0.2 percent of incident tritons in steady state. There are a wide variety of scientific questions that need to be addressed on this issue, including triton implantation, diffusion, and surface contamination in the pulsed, high-energy triton environment of an IFE wall with rapid thermal cycling. The tritium retention issue will also vary greatly with the choice of target-facing materials—for example, tritium can bond chemically with lithium.

Concerning liquid walls, they are separated into "thick," which implies that the energetic neutrons do not appreciably penetrate them (~50 cm), and "thin," in which the neutrons are not absorbed and thus hit the wall behind the thin liquid layer. Liquid gallium could be an excellent thin-wall material because it melts just above room temperature and has negligible vapor pressure even at very high temperatures. It would not, however, allow the necessary breeding of tritium if it were "thick,"—that is, the breeding ratio would be less than 1. Other materials that remedy this shortcoming are fluorine lithium beryllium (FLiBe), Pb, PbLi, and Li. All have vapor pressures that lead to a target chamber pressure of ~1 mTorr at a wall temperature of ~900 K, which is consistent with suitable flow and thermal transfer properties. In all cases, there need to be extensive studies of aerosol and vapor formation under conditions consistent with IFE shot conditions, so that it is confirmed that the target chamber can be cleared between shots at ~10 Hz.

FLiBe is a eutectic salt of LiF and BeF_2,[36] that produces tritium (mostly from ^6Li but also from ^7Li and ^9Be). In addition, the ^7Li and ^9Be soften the neutron energy spectrum through (n, 2n) reactions, which can help reduce neutron irradiation damage. For a wall thickness of 24 cm, FLiBe is expected to have a tritium-breeding ratio of unity, and the neutron flux is reduced by a factor of 10.[37] Its properties for tritium breeding are considered marginal, because the tritium (and other hydrogen isotopes) form hydrogen fluoride; thus, one must maintain chemical conditions that balance retention versus release of this highly reactive compound from the wall/blanket. (It is possible that the MoF_3 to MoF_6 redox buffer reaction can mitigate this release.[38]) Decomposition of FLiBe would lead to the production of fluorine and beryllium, both environmental hazards. In a fission reactor environment, it is known that FLiBe is not decomposed to a large extent by X-rays. This,

---

[36] A.R. Raffray and M. Zaghloul, 2002, "Scoping Study of FLiBe Evaporation and Condensation," Presentation at ARIES-IFE Project Meeting, General Atomics, San Diego, Cal., July 1-2; D.K. Sze and Z. Wang, 1998, "FLiBe—What Do We Know?," Presentation at the APEX/ALPS Project Meeting, Albuquerque, N. Mex., July 27-31.

[37] See C.L. Olson, 2005, "Z-Pinch Inertial Fusion Energy," *Landolt-Boernstein Handbook on Energy Technologies*, Volume VIII/3, Springer-Verlag, Berlin; G.E. Rochau and C.W. Morrow, 2004, "A Concept for a Z-Pinch Driven Fusion Power Plant," SAND2004-1180.

[38] Ibid.

however, needs to be confirmed in the more extreme conditions relevant to IFE. In this regard a question arises for the case where there is a magnetic field in the target chamber: FLiBe is a conductor, albeit a poor one, flowing in a magnetic field, so a voltage difference arises that could lead to electrolysis and hence the liberation of fluorine. In addition, relatively little is known about the extent to which FLiBe, Ga, etc., corrode the wall materials they coat, although use of vanadium alloys and ferritic steel is consistent with using FLiBe (particularly at the high temperatures envisioned for fusion chamber walls). One must also take into account the radioactive species produced by the neutrons, because these complicate routine operations and maintenance. For metals, many of these species have long half-lives of years; however, for FLiBe, although there are intense short-lived activities, most will decay quickly (in minutes and seconds).

No significant research at the appropriate engineering scale has been done on the hydrodynamic manipulation of these hot liquids to create the continuous wall coverage needed in a practical IFE plant. This means that large engineering facilities and their associated R&D programs will have to be brought into existence at an early stage for wet walls. In addition, there are obvious questions of cost and availability of Ga, Be, FLiBe, and the like in the quantities consistent with commercial-scale IFE.

The interaction of the high-energy neutrons with materials is not unlike that encountered in fission reactors, which has been studied for decades. The energies are, however, higher, and the dose rate dependence is likely to be quite different, as is the critical ratio of helium production to displacements. These neutrons both scatter and undergo nuclear reactions with atoms in the wall. These recoiling atoms and heavy reaction products create collision cascades of damage, which at the high wall temperatures coalesce into void and interstitial clusters. This can cause fundamental changes to materials (e.g., swelling).

Because the fusion neutron spectrum is much harder than that of fission, the fusion neutrons produce significantly more helium (10 to 1,000 times, depending on the material) in the bulk due to the (n, alpha) reactions. Because helium is insoluble in the materials, the accumulation of helium in voids and at grain boundaries can significantly degrade the material properties. The experience of fission is greatly limited in these effects due to its softer neutron spectrum. Over time, this damage leads to embrittlement, fatigue and other structural weakening. The (n, p) and (n, d) reactions produce hydrogen, which tends to migrate to grain boundaries and interstitial and defect sites. These effects were studied as part of the fast fission breeder program, in magnetic confinement fusion, and in ion implantation studies for semiconductor processing. To some extent, they can be investigated by using energetic heavy-ion beams, where the beam ions mimic the recoiling wall atoms. It is possible that total fluence data can be obtained in this way, but the effect of the very high dose rates will require special facilities.

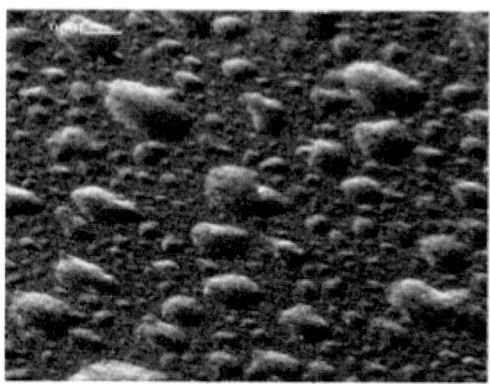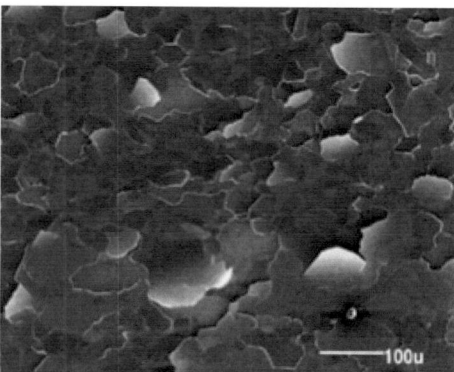

FIGURE 3.9 Examples of tungsten first wall materials damage due to ion implantation. SOURCE: Oak Ridge National Laboratory.

As mentioned earlier, the exposure of the wall surface to MeV and keV ions leads to recoil damage similar to that from neutrons, but it is much more localized. The consequence is sputtering of the surface, which changes its topography as material is removed. Just below the surface, the damage is intense, leading to blistering and exfoliation of wall material. Such effects have been studied; helium production is a major issue. Examples of first wall materials damage due to ion implantation are shown in Figure 3.9.

Although the final stages of the optical elements (mirrors or gratings) for a laser driver may be protected from ion damage by buffer gas and/or magnetic fields, their exposure to X-rays, ions, and energetic neutrons has to be addressed. Some work has been done in this area on fluence limits, but dose rate effects are not yet understood. In addition, where no buffer gas is present the effects from the accumulation of debris from the destruction of the target on the performance of these elements must also be considered.[39]

Path Forward

Most of the potential problems of the selection of appropriate materials for the walls and final stage optical elements have not yet been addressed at appropriate scale or under the appropriate environmental conditions. With this in mind, it is clear that a major research and development program with large-scale facilities is a necessary part of the development of IFE. It is the committee's belief that this

---

[39] L. Snead, N.M. Ghoniem, and J.D. Sethian, 2001, "Integrated Path for Materials R&D in Laser Inertial Fusion Energy (IFE)," Internal memorandum, Naval Research Laboratory, August.

program is of such a size and complexity that it should be structured very carefully. Its various parts need to be integrated with each particular IFE plant concept, because challenges are often specific to the details. Many materials issues involve understanding the basic science of materials interactions; research in these areas will benefit multiple designs. The timing of the R&D effort has to be matched to the schedule of milestones in the driver, target configuration, and chamber/wall design choices. Those portions that also occur in magnetic confinement fusion, such as neutron damage to structural materials, ion damage to first wall materials and tritium retention concerns, do not have to be duplicated, but one cannot assume that this research will automatically be relevant to both unless the program and facilities are designed with that dual use in mind. The choice of appropriate materials matters and must be considered an integral part of the roadmap to commercial IFE.

Since a decision about the choice of a specific IFE configuration has not yet been made, it is not feasible to suggest a detailed plan for the research and engineering associated with materials that extends in time out to the demonstration plant. A particular IFE configuration brings with it a special set of material-related issues to be addressed; thus, to address all possible materials problems ab initio would be both inefficient and expensive. For example, pulsed-power and heavy-ion fusion are not burdened by the issues of damage to final optical elements that hamper laser drivers. Direct-drive and indirect-drive laser IFE pose different challenges to wall materials, and solid and liquid walls are in themselves substantially different. Initial IFE materials R&D should focus on basic science issues common to multiple designs. The committee offers the following conclusion and recommendations.

Conclusion 3-13: Magnetic fusion energy (MFE) and inertial fusion energy (IFE) share the challenge of 14-MeV neutron damage, which cannot be addressed adequately by fission-reactor-based materials studies. Moreover, owing to the pulsed nature of IFE, there are critical differences between IFE and MFE in the capture and control of X-rays, energetic particles, and neutrons in the surrounding materials and their subsequent damage and response. IFE candidate material solutions will require some different testing and irradiation facilities.

Recommendation 3-5: When a particular IFE option is chosen, a materials R&D program focused on key technical issues should be established immediately and move in parallel with IFE development.

Recommendation 3-6: Since it may be important to identify obstacles in materials properties/performance in order to down-select among the various IFE options, it will be necessary to carry forward a modest materials program. This program should focus on issues that are common to the most

likely IFE choices and, in addition, try to anticipate the serious materials challenges that could affect the choice of an initial IFE prototype.

TRITIUM PRODUCTION, RECOVERY, AND MANAGEMENT

Background and Status

Tritium production, recovery, and management are key to the success of an IFE system. The supply of tritium on Earth is limited (half-life ~12.3 years), so tritium "breeding" is required to ensure a ready supply of fuel for IFE. Tritium self-sufficiency (the "closed" fuel cycle for fusion) is necessary for commercial success or even large-scale test facilities. This requirement brings with it a range of issues, including target performance, tritium breeding potential of the blanket, and the tritium inventory in the IFE system (because tritium is hazardous and readily mobile under certain conditions, there is a trade-off between tritium inventory and safety; see the section Environment, Health, and Safety Considerations, below).

The current section discusses the issues, challenges, and R&D surrounding IFE tritium production, recovery, and management. Several design studies have evaluated tritium-breeding performance and associated tritium inventories.[40,41] These studies provide a useful first examination of these issues. The quantitative conclusions from all such studies must be viewed as somewhat uncertain, because they are at a relatively high level and miss many of the issues that become apparent when a system is actually built at engineering scale, revealing, for example, the actual area available for tritium breeding once all the equipment, manifolding, and such are considered).

The tritium inventory in the target fabrication plant is highly dependent on the target performance (lower performance means higher tritium inventory in the targets) and on the process used for target fabrication (see Target Fabrication, above). Depending on the target fabrication process, tritium inventories in the target fabrication plant can be as large as 10 kg. Important in the consideration of tritium inventories is the ability to recover the unused tritium from the unburned DT fuel; as higher burn fraction results in less tritium to recover. In the LIFE concept, estimates suggest that about half of the tritium inventory will be in the target fabrication plant, and total tritium inventory in the LIFE system is <600 g.[42] The

---

[40] See the studies referenced in the previous section on OSIRIS, SOMBRERO, Prometheus-L and -H, HIBALL, HYLIFE, Z-Pinch, and LIFE.

[41] M. Dunne, E.I. Moses, P. Amendt, et al., 2011, Timely delivery of laser inertial fusion energy (LIFE), *Fusion Science and Technology* 60: 19-27.

[42] M. Dunne, R. Al-Ayat, T. Anklam, A. Bayramian, R. Deri, C. Keane, J. Latkowski, R. Miles, W. Meier, E. Moses, J. Post, S. Reyes, V. Roberts, LLNL, 2011, "Answers to the Second Request for Input from the NRC Committee on Prospects for Inertial Confinement Fusion Energy Systems," LLNL-MI-473693, Response to NAS IFE Committee questions.

SOMBRERO design study claims a similar (300 g) tritium inventory in the target fabrication plant.[43]

Tritium breeding is accomplished in the blanket. IFE and MFE share tritium breeding needs and basic blanket concepts. The section on reaction chambers above summarizes the types of chambers under consideration for IFE; they fall into two main categories: solid walls and liquid walls. Liquid lithium is an option for liquid walls and has the advantage of relatively high tritium solubility, thus reducing tritium permeation concerns; however, that high solubility can result in undesirably high tritium inventories. Tritium recovery systems have been partially developed and tested at laboratory scale,[44] and indicate that acceptable tritium removal and thus inventory limits can be met with these processes; further testing at laboratory and engineering scales is needed to confirm this. Liquid lithium is a superior tritium-breeding medium (compared with molten salt and LiPb) and is therefore attractive from a tritium self-sufficiency point of view.[45] Molten salt (e.g., FLiBe) and LiPb have the advantage from a safety point of view of reduced tritium inventories and less chemical activity; however, they have low tritium solubility and thus a higher driving force for permeation (a safety disadvantage) and may require tritium permeation barriers to control the movement of tritium throughout the system.

The SOMBRERO design, shown in Figure 3.10, is considerably different from most other IFE designs: it utilizes a granular Li_2O blanket (using gravity to move the particles through the system) that serves as both the coolant and the tritium breeder.[46] Low-pressure helium removes the tritium from the Li_2O and transports the granules to and from the intermediate heat exchangers. The tritium inventory in the SOMBRERO design was originally estimated at just under 200 g; however, later analysis indicated that the inventory may be 1-2 kg of tritium in the carbon structure,[47] emphasizing the potential for uptake of tritium in structural materials (see the section "Materials," above). A large tritium inventory requires an engineering or materials solution to ensure safety under off-normal conditions (see the section "Environment, Health, and Safety Considerations," below). Tritium removal from ceramic breeder blankets is also a topic of interest to the

---

[43] DOE, 1992, *OSIRIS and SOMBRERO Inertial Fusion Power Plant Designs*, DOE/ER-54100-1.
[44] Ibid.
[45] L. El-Guebaly and S. Malang, 2009, Toward the ultimate goal of tritium self-sufficiency: Technical issues and requirements imposed on ARIES advanced fusion power plants, *Fusion Engineering and Design* 84: 2072-2083.
[46] DOE, 1992, *OSIRIS and SOMBRERO Inertial Fusion Power Plant Designs*, DOE/ER-54100-1.
[47] G.L. Kulcinski, R.R. Peterson, L.J. Wittenberg, E.A. Mogahed, and I.N. Sviatoslavsky, 2000, "Dry Wall Chamber Issues for the SOMBRERO Laser Fusion Power Plant," UWFDM-1126, University of Wisconsin, Madison, June.

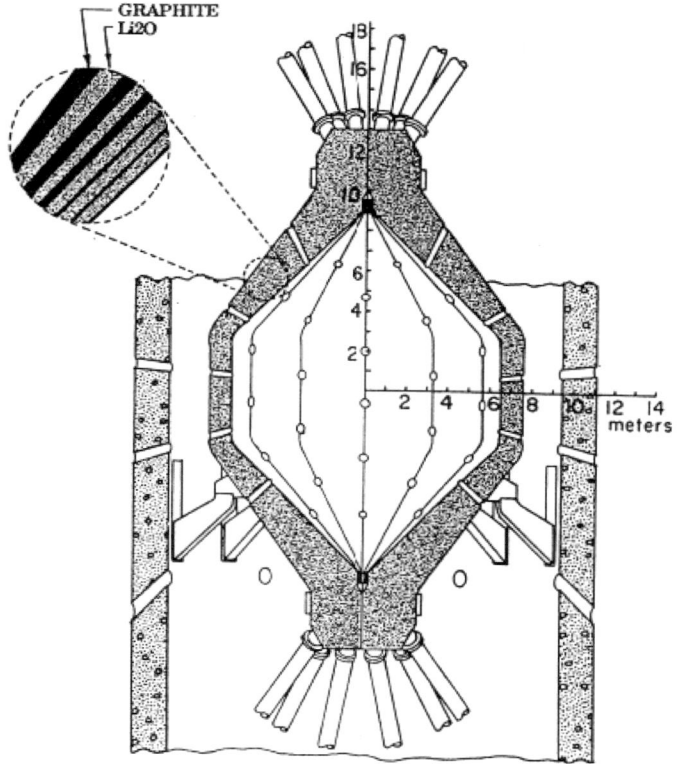

FIGURE 3.10 SOMBRERO's flowing Li_2O granule chamber concept. SOURCE: DOE, 1992, *OSIRIS and SOMBRERO Inertial Fusion Power Plant Designs,* DOE/ER-54100-1.

ITER test blanket module (TBM) program,[48] and the IFE program can leverage those activities.

Each of these studies shows tritium self-sufficiency. However, in actual application, losses due to uptake in structure, process losses, and actual neutron economy will likely be greater than accounted for in the studies. While there are a number of ways to adjust the tritium-breeding ratio (blanket thickness, $^6Li/^7Li$ ratio, neutron multiplier), until tritium breeding studies are done for detailed designs, including testing at engineering scale, the tritium self-sufficiency of any design must be considered uncertain. Tritium management will benefit from NIF and OMEGA studies to a limited extent (particularly target fabrication, tritium management,

---

[48] H. Albrecht and E. Hutte, 2000, Tritium recovery from an ITER ceramic test blanket module—Process options and critical R&D issues, *Fusion Engineering and Design* 49-50: 769-773.

tritium handling, and tritium processing). However, the lack of a breeding blanket in NIF leaves an important area in a state of uncertainty.

There has been limited work on liquid and solid breeder blankets in the IFE context. The work in the MFE program could be leveraged. Much could be gained from taking advantage of the larger MFE blanket programs under way in other countries.

Conclusion 3-14: Tritium-breeding performance has been considered in several design studies. These provide a useful initial examination of these issues. As these studies are at a preconceptual design level, they miss many of the issues that become apparent when a system is actually built at engineering scale.

Conclusion 3-15: Tritium recovery systems have been partially developed and tested at laboratory scale, and the signs are that acceptable tritium removal—and thus inventory limits—can be met with these processes. Further testing at laboratory and engineering scale is needed to confirm this.

Conclusion 3-16: Tritium management will benefit from National Ignition Facility (NIF) activities, particularly target fabrication, tritium management, tritium handling, and tritium processing. However, the lack of a breeding blanket in NIF leaves an important area uninvestigated.

Scientific and Engineering Challenges and Future R&D Priorities

The challenges associated with tritium production, recovery, and management are typically engineering and material challenges rather than fusion science challenges. More detailed designs are needed to reduce uncertainties in tritium production calculations. A better understanding of tritium permeation (and methods to reduce permeation) is needed, as is an understanding of tritium uptake in structural materials and tritium removal from breeding blankets.

Path Forward

Near Term (<5 Years)

Needed R&D activities include systems studies; tritium production and recovery studies in liquid and solid blankets (including predictive models); and target studies with a focus on increased burn fraction. Focus in the near term would be on modeling activities.

Medium Term (5-15 Years)

Success would be the validation of tritium production and recovery models in specially designed experiments. Testing of candidate thick liquid (and ceramic granules, if deemed promising in system studies) wall concepts in flow loops, including tritium extraction, and testing of candidate solid walls, including tritium extraction from coolant, would be carried out. Some new facilities would be needed.

Long Term (>15 Years)

The long-term objective would be to develop liquid-wall target chambers operating at 0.1 to 10 Hz or solid wall target chambers to be made available for an FTF and subsequent IFE demonstration plant.

Conclusion 3-18: More detailed designs are necessary to reduce uncertainties in tritium production calculations. A better understanding of tritium permeation (and methods to reduce permeation) is required, together with tritium uptake in structural materials and tritium removal from breeding blankets.

Recommendation 3-7: The work in the magnetic fusion energy program should be leveraged—in particular, the studies for the ITER Test Blanket Module program. Much could be gained from taking advantage of these larger MFE R&D programs under way in other countries.

ENVIRONMENT, HEALTH, AND SAFETY CONSIDERATIONS

Background and Status

Fusion energy has long been seen as having attractive environmental, health, and safety characteristics. The ability to separate the fuel (target) from the chamber system allows selection of structural materials that minimize the production of long-lived isotopes requiring long-term isolation (as is the case for used fuel from a fission reactor). From a safety perspective, tritium is one of the primary safety concerns, as it can be readily mobile under certain conditions. However the overall source term in a fusion system is small compared with the source term in a fission reactor; this should translate into advantages in licensing in the event that fusion approaches commercial deployment. Finally, consideration must be given to the risk of proliferation associated with IFE systems. The committee has had a companion committee, the Panel on the Assessment of Inertial Confinement Fusion Targets, whose charter calls for the consideration of proliferation issues as well as assessment of target physics, and it has included review of classified materials as needed. The final report of this panel includes its conclusions on proliferation issues related to energy applications of inertial fusion (see Appendix H).

The present section discusses the issues, challenges, and R&D needed to address environment, health, and safety considerations, including plant operation and maintenance, waste streams, and licensing and regulatory considerations.

Plant Operations and Maintenance

Because IFE plants will require a large capital investment, they are most suited for baseload operations. This will require minimal downtime, an attribute that has been attained by commercial fission plants in the United States (demonstrating over 90 percent availability on average), but only after many years of operational experience. The fission industry has developed a tightly coordinated set of maintenance activities that are timed to coincide with fueling outages; IFE plants should be able to develop a similar set of coordinated maintenance activities, but it will take some years of operational experience to do so.

Several design studies have included a discussion of maintenance.[49] Avoiding frequent replacement of components that are difficult to access and replace will be important to achieving high availability. Such components will need to achieve a very high level of operational reliability. Technology challenges discussed in this chapter must be overcome to maximize availability, and operating experience in prototypical environments is needed.

An important contributor to good availability is hands-on maintenance wherever possible.[50] Hands-on maintenance is typically faster than remote maintenance and may be less expensive.[51] Activation products in coolant streams should

---

[49] See, for example, M. Dunne, R. Al-Ayat, T. Anklam, A. Bayramian, R. Deri, C. Keane, J. Latkowski, R. Miles, W. Meier, E. Moses, J. Post, S. Reyes, and V. Roberts, LLNL, 2011, "Answers to the Second Request for Input from the NRC Committee on Prospects for Inertial Confinement Fusion Energy Systems," LLNL-MI-473693, Response to NAS IFE Committee questions; DOE, 1992, *OSIRIS and SOMBRERO Inertial Fusion Power Plant Designs*, DOE/ER-54100-1; DOE, 1992, *Inertial Fusion Energy Reactor Design Studies Prometheus-L and Prometheus-H*, DOE/ER-54101; B. Badger, K. Beckert, R. Bock, et al., 1981, *HIBALL—A Conceptual Heavy Ion Beam Fusion Reactor Study*, UWFDM-450, University of Wisconsin at Madison, and KFK-3202, Kernforschungszentrum Karlsruhe; J.A. Blink, W.J. Hogan, J. Hovingh, W.R. Meier, and J.H. Pitts, 1985, *The High Yield Lithium Injection Fusion Energy (HYLIFE) Reactor*, UCRL-53559, LLNL.

[50] M. Dunne, R. Al-Ayat, T. Anklam, A. Bayramian, R. Deri, C. Keane, J. Latkowski, R. Miles, W. Meier, E. Moses, J. Post, S. Reyes, V. Roberts, LLNL, 2011, "Answers to the Second Request for Input from the NRC Committee on Prospects for Inertial Confinement Fusion Energy Systems," LLNL-MI-473693, Response to NAS IFE Committee questions.

[51] S.J. Piet, S.J. Brereton, J.M. Perlado, Y. Seki, S. Tanaka, and M.T. Tobin, 1996, "Overview of Safety and Environmental Issues for Inertial Fusion Energy," INEL-96/00285.

be minimized to avoid exposure of plant personnel and maximize hands-on maintenance. Because fusion plants use tritium for fuel, maintenance activities must be done with attention to its presence (it can be very mobile; see the tritium management section above). Some designs utilize modular components for ease of maintenance and replacement.[52] Remote maintenance will be needed for some components and areas of the power plant. The IFE program should learn from remote maintenance activities at ITER and NIF and from the extensive long-term program on the Joint European Torus (JET).[53]

Because there are at present no important IFE test facilities that include a significant technology mission, there is no opportunity to test in IFE-prototypic conditions. As was discussed earlier in this section, achieving high levels of component reliability requires substantial testing and qualification of fusion components, far beyond what is available today.

The environment, health, and safety issues associated with plant operations and maintenance of an IFE power plant are expected to be substantially similar to those of current fission nuclear power plants. While fusion reactors will not have to deal with nuclear fuels and their resulting fission products, high levels of radiation and large amounts of radioactive materials will have to be safely handled. These will come from activation of the structural materials of the reactor and activated corrosion products in the coolant streams, as well as the presence of tritium. Fusion reactors will have to deal with significantly larger quantities of tritium than do fission reactors, as is discussed in the section "Tritium Production, Recovery, and Management," above.

Waste Streams

The IFE design studies that have been done over the years typically quantify the waste streams associated with each design.[54] The U.S. Nuclear Regulatory

---

[52] M. Dunne, R. Al-Ayat, T. Anklam, A. Bayramian, R. Deri, C. Keane, J. Latkowski, R. Miles, W. Meier, E. Moses, J. Post, S. Reyes, V. Roberts, LLNL, 2011, "Answers to the Second Request for Input from the NRC Committee on Prospects for Inertial Confinement Fusion Energy Systems," LLNL-MI-473693, Response to NAS IFE Committee questions.

[53] See http://tinyurl.com/c78oqfz for more information.

[54] DOE, 1992, *OSIRIS and SOMBRERO Inertial Fusion Power Plant Designs*, DOE/ER-54100-1; DOE, 1992, *Inertial Fusion Energy Reactor Design Studies Prometheus-L and Prometheus-H*, DOE/ER-54101; B. Badger, K. Beckert, R. Bock, et al., 1981, *HIBALL—A Conceptual Heavy Ion Beam Fusion Reactor Study*, UWFDM-450, University of Wisconsin at Madison, and KFK-3202, Kernforschungszentrum Karlsruhe; J.A. Blink, W.J. Hogan, J. Hovingh, W.R. Meier, and J.H. Pitts, 1985, *The High Yield Lithium Injection Fusion Energy (HYLIFE) Reactor*, UCRL-53559, LLNL; C.L. Olson, 2005, "Z-Pinch Inertial Fusion Energy," *Landolt-Boernstein Handbook on Energy Technologies*, Volume VIII/3, Springer-Verlag, Berlin; J.D. Sethian, D.G. Colombant, J.L. Giuliani, et al., 2010, The science and technologies for fusion energy with lasers and direct-drive targets, *IEEE Transactions on Plasma Science* 38(4): 690-703; M.

Commission (US NRC) governs disposal of radioactive waste in the United States; the regulations are covered in the U.S. Code of Federal Regulations, 10CFR61.[55] IFE and MFE designs have focused on the use of "low activation materials" that minimize the production of isotopes with long half-lives, with a goal of eliminating—or reducing as much as possible—waste that requires geologic disposal (of course the material must still function in its intended role, and this provides many challenges; see the section on materials issues above). Near-surface disposal (as opposed to geologic disposal) depends on specific activity limits (SALs), which are based on the direct gamma exposure from gamma-emitting radionuclides and the inhalation and ingestion of beta-emitting radionuclides. The radionuclides in 10CFR61 are for fission-based isotopes, but applying the same methodology produces SALs for fusion-based isotopes.[56]

Fusion design studies have included a focus on minimizing the production of waste requiring geologic disposal. This has been done through careful choice of materials—for example, by limiting Nb and Mo impurities in structural material,[57] by using SiC-based, low-activation materials,[58] or by possibly filtering out some radioactive elements from liquid wall materials. These actions typically increase the cost of the plant but reduce the cost of disposal into a mined geologic repository such as WIPP or the stalled Yucca Mountain. Also, recycling target material is helpful for minimizing waste.

The fusion community has been successful in designing power plants that meet the goal of reducing or even eliminating the production of high-level waste. However, the amount of low-level waste that requires disposal, albeit near-surface, is likely to be very large.[59] Figure 3.11 shows a comparison of waste volume for magnetic fusion

---

Dunne, E.I. Moses, P. Amendt, et al., 2011, Timely delivery of laser inertial fusion energy (LIFE), *Fusion Science and Technology* 60: 19-27; J.F. Latkowski, R.P. Abbott, S. Aceves, et al., 2011, Chamber design for the laser inertial fusion energy (LIFE) engine, *Fusion Science and Technology* 60:54-60; L. Cadwallader, L. and L.A. El Guebaly, 2011, Safety and environmental features, *Nuclear Energy Encyclopedia: Science, Technology, and Applications*, p. 413, Wiley & Sons; L.A. El-Guebaly, P. Wilson, and D. Paige, 2006. Evolution of clearance standards and implications for radwaste management of fusion power plants, *Fusion Science and Technology* 49: 62-73.

[55] Code of Federal Regulations, Title 10: Energy, Part 61—Licensing Requirements for Land Disposal of Radioactive Waste (Nuclear Regulatory Commission), Office of the Federal Register, National Archives and Records Administration, Revised as of January 1, 1991.

[56] E.T. Cheng, 2000, Waste management aspect of low activation materials, *Fusion Engineering and Design* 48: 455-465.

[57] L.A. El-Guebaly and the ARIES Team, 2000, "Views on Neutronics and Activation Issues Facing Liquid-Protected IFE Chambers, Topical on Fusion Energy," 14th ANS Topical Meeting on Fusion Energy, Park City, Utah, October.

[58] L.A. El-Guebaly, P. Wilson, D. Henderson, L. Waganer, and R. Raffray, 2003, Radiological issues for thin liquid walls of ARIES IFE study, *Fusion Science and Technology* 44: 405-409.

[59] S. Reyes, J. Sanz, J. Latkowski, 2002, Use of Clearance Indexes to Assess Waste Disposal Issues for the HYLIFE-II Inertial Fusion Energy Power Plant Design, UCRL-JC-147039, LLNL, January 17, 2002.

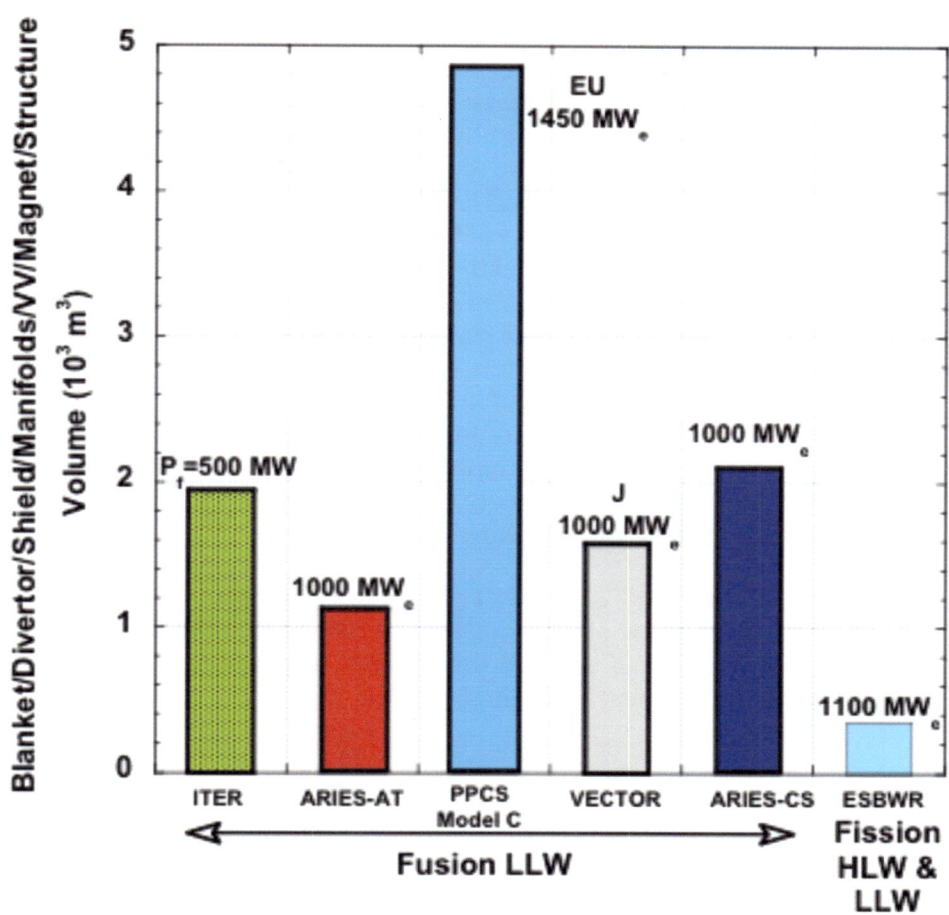

FIGURE 3.11 Lifetime radioactive waste volume comparison for various MFE designs (actual volumes of components; not compacted, no replacements; bioshield excluded). LLW, low-level waste; HLW, high-level waste. SOURCE: L. El-Guebaly et al., 2008, Goals, challenges, and successes of managing fusion activated materials, *Fusion Engineering and Design* 83: 928-935.

designs;[60] inertial fusion designs have similar volumes.[61] Low-level waste disposal facilities in the United States are becoming oversubscribed, and siting a new low-level waste disposal facility is also likely to be a very difficult. There have been some studies looking at the potential for recycling radioactive materials to reduce the amount

---

[60] L. El-Guebaly, V. Massaut, K. Tobita, L. Cadwallader, 2008, Goals, challenges, and successes of managing fusion activated materials, *Fusion Engineering and Design* 83: 928-935.

[61] S. Reyes, J. Sanz, J. Latkowski, 2002, Use of Clearance Indexes to Assess Waste Disposal Issues for the HYLIFE-II Inertial Fusion Energy Power Plant Design, UCRL-JC-147039, LLNL, January 17, 2002.

of material that must be stored.[62] Further analysis will be needed to determine the practicality and net cost of this approach. Recycling and reuse of materials within the fusion system—as opposed to "free release" of recycled material—is likely to meet with less resistance from regulators, the recycling industry, and the public.[63]

Of particular importance are those waste streams that are considered "mixed waste." Mixed waste has both a chemical hazard and a radiation hazard; irradiated lead is an example of a mixed waste. Lead is a coolant candidate as well as a target material candidate. Mixed waste currently has no disposition path in the United States, but regulations governing its disposal are under development and would likely be in place before deployment of the first commercial fusion plant.

Conclusion 3-19: Design studies of inertial fusion energy power plants indicate that, with the use of low-activation materials, it will be possible to minimize high-level waste. However, the amount of waste that requires disposal, albeit near the surface, may be very large. Low-level waste disposal in the United States is becoming increasingly difficult.

Recommendation 3-8: There have been studies that examine the potential for recycling and reuse of radioactive materials within the fusion system to reduce the amount of material that must be disposed of; the committee encourages the continuation of these studies.

Licensing and Regulatory Considerations

The United States Nuclear Regulatory Commission (NRC) is a conservative body. This is appropriate given its role in the oversight of U.S. commercial nuclear facilities. The vast majority of the NRC's licensing experience has been with light water reactors (LWRs), and its regulations, for the most part, have grown out of the LWR experience. Licensing a fusion power plant will require blazing new trails, and it will be important for the fusion community to work with the NRC to help it to understand the hazards (which are much different from the hazards in an LWR) and their mitigation in a fusion power plant. Communication early in the process is important to a successful outcome.[64]

---

[62] L. El-Guebaly, R. Pampin, M. Zucchetti, 2007, Clearance considerations for slightly-irradiated components of fusion power plants, *Nuclear Fusion* 47(7): S480-S484; L. El-Guebaly, M. Zucchetti, L.D. Pace, B.N. Kolbasov, V. Massaut, R. Pampin, et al., 2009, An integrated approach to the back-end of the fusion materials cycle, *Fusion Science and Technology* 52(2): 109-139.

[63] National Research Council, 2002, *The Disposition Dilemma: Controlling the Release of Solid Materials from Nuclear Regulatory Commission-licensed Facilities*, National Academy Press, Washington, D.C.

[64] R.A. Meserve, Carnegie Institution for Science, "Licensing a Commercial Inertial Confinement Fusion Energy Facility," Presentation to the Committee on October 31, 2011.

Some licensing and regulatory-related work has been done for the ITER program, and much of that work provides insights into IFE licensing processes and issues. The LIFE program has considered licensing issues more than any other IFE program; however, much more effort would be needed if IFE were to seriously pursue an NRC license. The Next Generation Nuclear Plant (NGNP) fission reactor project plans to license and build a high-temperature gas fission reactor. Gas reactors have been built and operated previously in the United States and Europe, although at lower operating temperatures than are envisioned for the NGNP. The licensing strategy developed for the NGNP provides a good picture of the challenges associated with licensing a relatively standard technology.[65]

The licensing of fission power plants is moving toward a risk-informed approach, whereas in the past it took primarily a deterministic approach. The LIFE program is developing a similar approach.[66] The favorable safety characteristics of the IFE and MFE fusion plants should simplify the licensing process; however, the burden of proof for IFE plants will be no different than for fission plants. One of the safety-related goals for fusion is to demonstrate that there would never be a need for public evacuation under any event. This is a clear example of the favorable safety characteristics of a fusion plant.

> **Conclusion 3-20: Some licensing/regulatory-related research has been carried out for the ITER (magnetic fusion energy) program, and much of that work provides insights into the licensing process and issues for inertial fusion energy. The laser inertial fusion energy (LIFE) program at Lawrence Livermore National Laboratory has considered licensing issues more than any other IFE approach; however, much more effort would be required when a Nuclear Regulatory Commission license is pursued for inertial fusion energy.**

Safety analysis has been an important part of the IFE design studies cited earlier. Early analyses were relatively simple. They often looked at total inventories of radioactive material and determining how much material could be released based on total system energy. These analyses have given way to more sophisticated analyses, sometimes employing tools originally developed for the fission industry and adapted to fusion.[67] Tritium inventory and release mitigation is an important part of the fusion safety case. Tritium can be highly mobile under certain condi-

---

[65] Next Generation Nuclear Plant Licensing Strategy—A Report to Congress, www.ne.doe.gov/pdfFiles/NGNP_report toCongress.pdf, August 2008.

[66] M. Dunne, E.I. Moses, P. Amendt, et al., 2011, Timely delivery of laser inertial fusion energy (LIFE), *Fusion Science and Technology* 60: 19-27.

[67] B.J. Merrill, A lithium-air reaction model for the MELCOR code for analyzing lithium fires in fusion reactors, *Fusion Engineering and Design* 54: 485-493.

tions, so minimizing its inventory in fusion facilities is a first step (see the section on tritium management above). Other radioactive material present in the IFE plant must also be considered, together with possible release scenarios. Overall, the IFE source term is significantly smaller than its fission counterpart, which should benefit the licensing process. Analysis done for systems studies shows acceptable safety performance; however, in the absence of experimental results to validate models, the actual performance remains highly uncertain. Validation and verification of models is extremely important to the NRC and will be an important factor in the licensing process.

Recommendation 3-9: Validation and verification of models is extremely important to the Nuclear Regulatory Commission and will be an important factor in the licensing process. Development of models, including validation and verification, should be pursued early. Working with the NRC early and often will be important, as well as looking to other programs (e.g., ITER and fission) for successful licensing strategies.

Scientific and Engineering Challenges and Future R&D Objectives

The environmental, safety, and health aspects of the IFE facilities should continue to be an important point of discussion in any program. The IFE community should continue to analyze and bring attention to the favorable characteristics of these plants. Continued development of sophisticated models, together with data for validation of the models, is important in the preparation for licensing of an IFE plant. The IFE program should continue to keep abreast of NRC licensing activities and keep the lines of communication with the NRC open.

Path Forward

Near Term (<5 Years)

Needed R&D activities include systems studies with a focus on realistic assumptions and schedules. Radioactive waste management should be an area of particular focus given recent activities by the Blue Ribbon Commission on America's Nuclear Future (BRC).[68] The development of a safety model, with an eye towards future licensing, and the development of experiments to validate models will be critical.

---

[68] The BRC was created under the authority of DOE and tasked with devising a new strategy for managing the back end of the nation's inventory of nuclear fuel cyclewaste; it issued its final report in January 2012. A copy of the report and other information on the commission can be obtained at http://tinyurl.com/bvsshko; accessed on May 16, 2013.

Medium Term (5-15 Years)

Concepts for recycling IFE target and chamber materials need to be studied experimentally, possibly using only nonradioactive elements. Experiments would be done to benchmark accident analysis codes with materials and configurations typical of fusion power plant designs. Success would be experimental validation of safety models.

Long Term (>15 Years)

The long-term objective would be to begin development of the licensing case for an IFE demonstration plant.

BALANCE-OF-PLANT CONSIDERATIONS

The purpose of an IFE power plant is to produce useful energy in the form of electricity or high-temperature process heat, or chemical energy in the form of hydrogen. To do this, the power plant must convert the energetic products of fusion reactions—high-energy neutrons and charged particles—into the desired useful forms. To become a practical source of energy, IFE must produce and convert the fusion energy in a manner that is technically feasible, environmentally acceptable, and economically attractive compared to other long-term, sustainable sources of energy.

The high-energy neutrons and charged particles from the fusion reactions deposit their thermal energy on the walls of the reaction chamber and in the tritium-breeding blanket surrounding the chamber. Everything outside the chamber and blanket, excluding the laser or particle beam drivers or the pulsed power system, is considered the "balance of plant" (BOP). The BOP includes the systems for conversion of thermal energy to electricity, the buildings and structures for the power plant, and all the conventional services. While schemes have been proposed to convert some of the charged-particle energy directly into electricity by electrostatic or magnetohydrodynamic processes, first-generation IFE power plants will most likely utilize fairly conventional thermal power conversion systems to convert the energy contained in the hot coolant from the chamber wall and blanket into electricity. Similar "heat engine" thermal power conversion systems are widely used on nuclear fission power plants and on fossil-fired power plants around the world. The Rankine cycle, or steam cycle, and the Brayton cycle, or gas-turbine cycle, are widely used heat engines that appear well suited for application to the conversion of thermal energy from fusion into electricity. There appears to be little need for power conversion system development that would be unique to fusion or IFE, although IFE-specific BOP designs will need to be developed, and opportunities for innovation should always be welcome.

Conclusion 3-21: Existing balance-of-plant technologies should be suitable for IFE power plants.

The thermal conditions—inlet and outlet coolant temperatures—proposed for IFE power plants are similar to those used by fission and fossil power plants today, so that the BOP for an IFE power plant should likewise be very similar to those used today. An area of concern is that of system interfaces and the possibility of hazardous material transport across those interfaces. The IFE reaction chamber will contain quantities of radioactive tritium, radioactive target debris, and some radioactive material sputtered from the first wall. In addition, it will operate at elevated temperatures. Tritium may migrate through the chamber walls and into the primary coolant stream. The coolant will pass through heat exchangers, and tritium may migrate through the heat exchangers into the secondary coolant and eventually into the rest of the power plant and even into the environment. These issues are part of the larger tritium control issue discussed in the section on tritium management, above. These interface concerns may require R&D to develop coatings for BOP components and heat exchangers that are resistant to permeation by tritium and tritium removal systems for the various chamber, blanket, and power conversion system coolants.

Path Forward

Near Term (<5 Years)

The design and analysis of BOP systems will continue to be included in IFE system studies and design studies, with emphasis on identification and evaluation of critical issues.

Medium Term (5-15 Years)

As favored design concepts begin to emerge, R&D into critical issues that have been identified—such as tritium permeation and control—will need to be carried out.

Long Term (>15 Years)

BOP systems will need to be developed and deployed as part of demonstration IFE systems.

ECONOMIC CONSIDERATIONS

An essential requirement for any new energy system to compete in future markets is to offer a product at a competitive price. For an IFE power plant, the main measure is the cost of electricity (COE). The formula for the COE is typically given by:

$$COE = (C_{cap} \times FCR + C_{fuel} + COM)/(P_{enet} \times 8{,}760 \text{ (hr)} \times F_{cap}) + Decom$$

where C_{cap}, construction costs including interest charges during construction; FCR, fixed charge rate; C_{fuel}, fuel costs including targets; COM, operations and maintenance; P_{enet}, net electric power; F_{cap}, capacity factor; and Decom, annual decommissioning charge in mills per kilowatt-hour or $/MWh, which can be calculated as the cost of decommissioning, times the appropriate annual sinking fund factor to accumulate those funds, divided by the amount of electricity produced per year ($P_{enet} \times 8{,}760$ (hr) $\times F_{cap}$).

Conclusion 3-22: An essential requirement for any new energy system to compete in future markets is to offer a product at a competitive price. For an IFE power plant, the main measures are the cost of electricity generation and, in particular, the capital cost.

The capacity (or sometimes called the availability) factor (F_{cap}) has a large influence on the COE. It is the crucial number in converting capital costs to COE. IFE power systems will be very capital-intensive systems with perhaps relatively modest fuel costs, provided the goals of low-cost targets can be met (discussed further below). Such plants will likely operate as base-load power plants where a premium is placed on operating at the maximum capacity factor. IFE power plant studies typically assign a value of 70 percent to 80 percent to F_{cap}. These values cannot be achieved today given the early stages of IFE technology development, so really they represent a goal. By way of comparison, the current fleet of fission power plants in the United States routinely achieves an average capacity factor of about 90 percent.

Achieving high capacity factors requires two basic features of the system: high component reliability (usually measured by the mean-time-to-failure for each component) and acceptable maintenance or downtimes (usually measured by the mean-time-to-repair for each component). There is a strong relationship between the allowed values of the mean-time-to-failure and the mean-time-to-repair for a given component. The longer mean-time-to-repair, the longer must be the mean-time-to-failure. In other words, the harder it will be to replace the component, the higher must be the degree of reliability. Defining the acceptable values for the mean-time-to-failure and mean-time-to-repair for all the components in a complex IFE power plant will require a comprehensive systems engineering approach.

Achieving high levels of component reliability requires substantial testing and qualification of fusion components, far beyond what is available today. For example, no fusion reaction chamber has ever been built and certainly none has been tested to the extent needed to establish failure modes and a reliability database. Given the large number of components and systems in an IFE power plant (and an MFE power plant), a substantial investment of time and money will be required to conduct those tests and they will have an enormous impact on the overall time horizon for developing commercial IFE systems. Although much useful testing can and will be done in simulation facilities, at some time, testing in an actual fusion environment will be needed. These very large investments with long timescales will thus have a profound impact on the roadmap for realizing fusion power systems. While ITER and a future IFE demonstration plant are very different, it should be possible to take advantage of some of the experience with ITER—for example, the hardware and procedures developed for remote maintenance—to reduce the implementation time for an IFE demonstration plant.

Achieving high capacity factors (availability) in light of an IFE system's components is an equally challenging task. Some of these components will necessitate using remote handling systems. While the technology and experience in other fields (e.g., fission reactors and space systems) can be adapted to fusion needs, there exists today very limited experience with remote maintenance in fusion systems. ITER is one very important source of such information. Developing the maintenance systems for an IFE power plant will entail a significant effort, but there is very little work under way today in the United States to support those efforts.

For these reasons, the capacity factor is probably the most unpredictable of all the factors that affect the COE. This is true of both fusion concepts, inertial and magnetic.

Conclusion 3-23: As presently understood, an inertial fusion energy power plant would have a high capital cost and would therefore have to operate with a high availability. Achieving high availabilities is a major challenge for fusion energy systems. It would involve substantial testing of IFE plant components and the development of sophisticated remote maintenance approaches.

Of special concern for the economics of IFE is the cost of the targets. The feasibility of developing successful fabrication and injection methodologies at the low cost required for energy production—about $0.25 to $0.30/target,[69] or about

---

[69] W.S. Rickman and D.T. Goodin, 2003, Cost modeling for fabrication of direct drive inertial fusion energy targets, *Fusion Science and Technology* 43(3): 353-358.

one ten thousandth of current costs, and at a production rate that is 100,000 times faster than current rates—is a critical issue for inertial fusion. IFE researchers working on target capsule costs argue that between increased yields and batch-size increases, cost reductions of two orders of magnitude are possible with significant development programs.[70] It appears that the target-cost numbers may be possible, although challenging, considering the number of assumptions and judgments that are needed to get to the desired reduction of a factor of 10,000.

Conclusion 3-24: The cost of targets has a major impact on the economics of inertial fusion energy power plants. Very large extrapolations are required from the current state-of-the-art for fabricating targets for inertial confinement fusion research to the ability to mass-produce inexpensive targets for inertial fusion energy systems.

Construction or capital costs are typically divided into fusion-specific components (e.g., laser or particle-beam drivers, chambers, and target fabrication and injection) and the BOP. The BOP was discussed in the preceding section and will likely rely on existing concepts with cost estimates that are relatively well known. Cost estimates for the fusion components necessarily entail more uncertainty because in some instances (e.g., chambers and high-capacity target fabrication) they are still in the very early stages of development. Nevertheless, the construction costs have less uncertainty than the capacity factor.

In fission electricity experience, standard project costs (e.g., owner's cost and engineering during construction) are typically taken as a percentage of the basic capital cost. Escalation and inflation factors may also be incorporated.

The IFE COE estimated in various studies falls between 5 and 10 cents/kWh in current dollars.[71] These estimated COEs for IFE power plants are in the same general range as COEs for other energy options, but because of the relatively early phase of the development of IFE components and systems, much uncertainty surrounds them. It appears that the COE numbers obtained in past studies are pos-

---

[70] D.T. Goodin, N.B. Alexander, L.C. Brown, D.T. Frey, R. Gallix, C.R. Gibson, et al., 2004, A cost-effective target supply for inertial fusion energy, *Nuclear Fusion* 44(12): S254-265.

[71] DOE, 1992, *OSIRIS and SOMBRERO Inertial Fusion Power Plant Designs*, DOE/ER-54100-1, Volume 1. Executive Summary and Overview; T. Anklam, LLNL, "Life Delivery Plan," Presentation to committee on March 30, 2011; B. Badger, D. Bruggink, P. Cousseau, et al., 1995, *LIBRA-SP, A Light Ion Fusion Power Reactor Design Study Utilizing a Self-Pinched Mode of Ion Propagation*—Report for the period ending June 30, UWFDM-982 University of Wisconsin Fusion Technology Institute; J.T. Cook, G.E.Rochau, B.B. Cipiti, C.W. Morrow, S.B. Rodriguez, C.O. Farnum, et al., 2006, *Z-Inertial Fusion Energy: Power Plant*, SAND2006-7148, SNL; M. Dunne, LLNL, "Overview of the LIFE Power Plant," Presentation to the committee on January 29, 2011; I.N. Sviatoslavsky, et al., 1993, "SIRIUS-P, An Inertially Confined Direct Drive Laser Fusion Power Reactor," UWFDM-950, University of Wisconsin Fusion Technology Institute.

sible, but they contain uncertain components owing to the untested assumptions that must be made when making estimates for new technology.

Financing and business considerations, such as the fixed charge rate (capital charge rate), will have an important influence on the COE. Usually this is made up of two parts: a charge rate for the share held by equity investors and a (lower) charge rate for the debt-investor share. These terms can vary based on the confidence investors have in the readiness and cost-effectiveness of the technology and the extent to which the investment is protected. Investment can be protected in some states by a decision of the public utility commission. Debt investment can be protected by federal loan guarantees or by direct federal assumption of the debt. The charge rate for IFE will be determined by the entire history of the technology. The more complex the technology, the more prone it is to delays and bumps along the road to implementation and the bigger the effect on investor and guarantor psychology.

For example, most past IFE cost of electricity studies did not carry individual uncertainty ranges. Some of the difficulties in using estimates of electricity costs for IFE in comparison with other energy technologies or among IFE options could be overcome, in part, if uncertainty ranges were a required component of cost estimates.

It is not clear to what extent the COE studies for IFE are "forward" estimates (made without looking at a cost goal) or "backward" estimates (made with an eye on a cost goal), or a mixture of the two. Certainly, the BOP estimates can be based on conventional databases of cost elements and would qualify as forward cost estimates. They can be compared to cost estimates made for other, traditional energy technologies, with the caveat that future estimates for all technologies may be low when compared to actual as-built and as-operated facilities. Hence, cost estimates for fusion, even were they to be based totally on forward calculations, should be compared to estimates of future COEs for other technologies, not current-day market prices.

Cost estimates for the purely fusion components of the COE may have been, to some degree, backward estimates, starting from values based on views of future prices of the alternatives. Analysts taking this approach would determine if it was possible to reach such targets for the fusion components of the COE and then use those possible numbers to compute a total COE. In such cases, the fusion COEs might be better labeled "possible values" rather than COE estimates.

In addition to predicting possible COE values, cost analysis can help to identify where R&D dollars should be targeted. The sensitivity of total cost-to-cost variations in system components helps to identify where a reduction in cost (via R&D, for example) would have the greatest impact. The effectiveness of such analyses depends critically on having a well-developed system engineering capability.

Similarly, the technology readiness level (TRL) process is another useful tool that is also discussed in Chapter 4.[72] In dealing with uncertainty ranges, the use of TRLs for each component, with separate uncertainty ranges on the component COE appropriate for different TRLs, could help planners decide on where in order to allocate resources to lower costs. Such a methodology would help to standardize cost and uncertainty estimates across different fusion technologies; it is discussed further in Chapter 4.

Use of TRLs and other readiness concepts, such as "integration readiness levels,"[73] also provide structure for keeping costs under control. There have been problems historically with cost escalation in government/industry partnerships from which useful lessons for IFE can be drawn. For instance, many large DOE programs/projects did not proceed as planned. Although there are many reasons why projects may fail technically or not meet their cost objectives, two stand out and are worth special consideration given the charge to this committee: (1) the breakdown of large, multiowner projects, and (2) significant cost increases in large, first-of-a-kind demonstration or prototype plants. The committee believes that the TRL methodology should be required for all major components of the IFE program.

It is important to note that the COE for IFE may not be the most immediate obstacle to successful development. At the size currently envisioned in most studies, the total cost of an IFE plant may be the biggest obstacle to IFE development, when looked at through the prism of current-day electricity company concerns. Given the rapid escalation in capital costs over the last decade, projected costs of gigawatt facilities for all capital-intensive electricity plants have reached the point where they represent a significant fraction of company capitalizations, making investments a "bet-the-company" decision. Efforts are under way to downsize electricity plants to reduce the sticker shock. A national IFE program should explore a range of plant sizes given the uncertain market and financial situation in this country in the coming decades. In particular, it is very important to understand the lower bound for an IFE plant output in terms of key physics constraints (e.g., target energy gain) and engineering constraints.

Conclusion 3-25: The financing of large, capital-intensive energy options such as an IFE power plant will be a major challenge.

R&D can attempt to address the two major economic obstacles confronting IFE—namely, skepticism about reaching cost/kWh targets and the high cost per

---

[72] DOE, 2011, *Technology Readiness Assessment Guide*, DOE G 413.3-4A, Washington D.C.: Department of Energy.
[73] See J.C. Mankins, 2002, Approaches to strategic research and technology (R&T) analysis and road mapping, *Acta Astronautica* 51(1-9): 3-21 and B. Sauser, J.E. Ramirez-Marquez, R. Magnaye, and W. Tan, 2008, A systems approach to expanding the technology readiness level within defense acquisition, *International Journal of Defense Acquisition Management* 1: 39-58.

plant. R&D can also attempt to reduce investor risk, whether for government or private investors, by encouraging innovation in IFE components and designs, improving TRLs through engineering advances, and by laying the ground for spin-offs of private companies.

Systems analysis—in this context, the purely technical quantitative assessment of the expected performance of various interconnected technologies—is an important tool in the development of any complex system.[74] Systems analysis can also identify outcomes of various implementation scenarios based on various assumptions. It is primarily concerned with the performance of various technologies and does not address the pathways or nontechnical constraints in achieving the implementation of those technologies. However, it does enable assessing the sensitivity of the system to nontechnical constraints translated into system impacts. Cost assessment is one of the outcomes of a systems analysis, as discussed earlier.

As already mentioned, the cost of a plant generating 1 GW or more of electricity represents a considerable portion of the book value of any U.S. company likely to build a fusion reactor: this is in and of itself a huge barrier to entry. There is another problem specific to those high-capitalization facilities that might be built in the many states in the United States in which competitive, short-term electricity markets have been established. A fusion facility, like a nuclear fission facility, will not pay off its investors for a long time. In the absence of long-term contracts, these facilities would endure an extended period of vulnerability to market prices dropping, forcing bankruptcy and massive losses. While it could be that long-term contracts will be established in such markets in the years ahead, until that time, investments in expensive, capital-intensive projects are risky in competitive markets. Investors would therefore be looking for a high rate of return, driving up the per-kilowatt-hour cost.

The fission industry is working to modularize and downsize electricity plants to reduce the high costs and impact on the grid. Fusion R&D might want to follow that example. One goal of R&D could be to design IFE power plants that are naturally smaller or radically cheaper or to improve existing designs. Designers might explore modular systems in which relatively small fusion devices—built in sequence as finances allow—share common driver facilities. The assignment of an "investor readiness level" to a design, including differentiated levels of readiness to venture capitalists, equity investors, and debt investors, could prove a useful discipline for planning. Even though the COE might be higher, a smaller plant might be more viable in the United States because its total cost is more attractive to potential investors.

---

[74] K.A. McCarthy and K.O. Pasamehmetoglu, "Using Systems Analysis to Guide Fuel Cycle Development" (Paper 9477, INL/CON-09-15764). In: *Global 2009*, Paris, 2009.

Because it is not possible to anticipate which business model will be the most successful decades from now, a long-range technology should have an eye on supporting multiple business models. These models range from those in which the U.S. government stands behind the technology, maintains a high percentage of the ownership of the construction, and even operates the plant, to a model in which venture capitalists support small companies and obtain key patents on IFE components, to a model where the government builds a few facilities with the idea that private companies will step in afterward to improve and market the by then proven technology.

Government support for R&D, as part of or in addition to systematic engineering approaches, could greatly benefit IFE under all of these business models. Rewarding innovation as part of engineering could provide a stronger base from which spinoff companies could arise. Encouraging ideas from a larger community than is now involved in IFE efforts could contribute to increased innovation and could also increase the number of patents likely to be developed, which is a prerequisite for the venture capital model.

Based on the information in this section and its conclusions, the committee makes three recommendations:

Recommendation 3-10: Economic analyses of inertial fusion energy power systems should be an integral part of national program planning efforts, particularly as more cost data become available.

Recommendation 3-11: A comprehensive systems engineering approach should be used to assess the performance of IFE systems. Such analysis should also include the use of a technology readiness levels (TRLs) methodology to help guide the allocation of R&D funds.

Recommendation 3-12: Further efforts are needed to explore how best to minimize the capital cost of IFE power plants even if this means some increase in the cost of electricity. Innovation will be a critical aspect of this effort. Possible options include use of a smaller fusion module, even at higher specific capital cost per megawatt of electricity, and the use of a fusion module for which capital cost is reduced by accepting a higher operating cost.

4

A Roadmap for Inertial Fusion Energy

The statement of task for this study charged the committee to "advise the U.S. Department of Energy on its development of an R&D roadmap aimed at creating a conceptual design for an inertial fusion energy demonstration plant." While crucial milestones such as ignition and reactor-scale gain have yet to be achieved, the committee judges that inertial fusion energy (IFE) has made sufficient progress that a roadmap can be usefully considered as part of planning for an IFE segment of the long-term U.S. energy portfolio (see Conclusion 1-1). This chapter will consider the status of the options that are discussed in the previous chapters and will develop an approach for a composite event-based roadmap.

The committee had extensive discussions about which type of roadmap for an IFE demonstration plant would best meet the needs of the Department of Energy (DOE) and its oversight committees and agencies. The classical approach to roadmapping is to develop time-based phases and budgetary levels required to complete each phase. The main advantage of this approach is that a timeline is set and the needed resources are delineated. However, for IFE, uncertainties in the pace of scientific understanding and technology development—and the vagaries of the budgeting process—make it difficult, if not impossible, to maintain a time-based roadmap. Thus, the committee decided that a milestone-based (in other words, event-based) roadmap would be most appropriate here.

In this chapter, the committee sets out the roadmapping approach that best fits the needs of DOE; considers the status of development of the IFE options (i.e., laser-, ion beams-, pulsed power-based, etc.); lists the critical milestones that each of the options must reach in order for development of that option to continue;

and then constructs the first element of an event-based roadmap—that portion leading to ignition. It also lays out a conceptual path for steps leading to success: i.e., the decision to proceed with the conceptual design of a demonstration plant (DEMO). A discussion of key terminology leading to a DEMO is given in Box 4.1.

The DEMO, which will test many technologies together at or near full scale for the first time, will not be expected to work flawlessly as designed or even economically in its early stages. In fact, the DEMO should be designed for ease of retrofitting, and it will have extensive monitoring capabilities, which will increase

---

BOX 4.1
Description of Programmatic Terms Used in This Chapter

The committee decided that a milestone- or event-based roadmap is most appropriate for IFE because of the current stage of technical maturity. However, before describing this road mapping approach, a few definitions are needed.

- Technology Application (TA). The committee has defined a technology application as a combination of a driver-target-chamber approach that has been discussed in the previous chapters and is included in this road mapping exercise because of its potential for success, scientific results to date, and level of development. For simplicity, we define three TAs based on the three main driver approaches: lasers, heavy ions, and pulsed-power.
- Integrated Research Experiment (IRE): An IRE tests the simultaneous operation of several aspects of a fusion reactor, but not necessarily all of them. For example, a single laser driver module would be aimed at injected surrogate targets at a rate of up to a reactor's repetition rate to test driver quality, target launching, tracking, and interception. Such facilities might be upgraded to include a few modules, for example, for undertaking scaled implosions for speeding up the testing of targets. For pulsed power, the equivalent would be demonstrating repetitive recyclable-transmission-line replacement at high power without arcing.
- Fusion Test Facility (FTF): The FTF is a demonstration of repetitive deuterium-tritium (DT) target shots using reactor-scale driver energy that generates high gain for the relevant TA. An FTF may be used initially for demonstrations of gain at very low frequency, followed by an increasing repetition rate to within an order of magnitude of the repetition rate of a commercial power plant, accumulating a total number of shots exceeding, say, 10^6 per year, or perhaps 10^5 for pulsed power fusion (since pulsed-power would operate at a lower repetition rate and higher yield/target compared to other approaches). As experience is gained with a successful TA, the FTF might be used to accumulate operating experience with longer run times.
- Demonstration reactor (DEMO): A demonstration reactor has to deliver enough electric power to the grid over 5 to 10 years to enable industry to judge the potential commercial viability of IFE through the conduct of reliability analyses, to establish reasonable cost estimates, and to assess safety sufficiently well to ensure that commitments could be made for construction and economical operation of commercial fusion power plants that must operate for more than 25 years.

its capital costs. Nevertheless, the DEMO will be built when technology is at such a level that a successful DEMO could provide the confidence needed for the private sector to take on IFE as a commercial product, albeit with some design modifications and some initial government assistance. There is a continuum of technology levels between a Fusion Test Facility (FTF) and a DEMO, so a sufficiently complete set of driver, target, and chamber data leading straight to an early DEMO, by-passing an FTF, is not precluded but is highly unlikely.

In addition, assuming that progress in one or more approaches to practical IFE can be realized, the organizational structure for conducting the research must be considered as well as the potential program costs. However, since IFE research is currently funded only at a low level and in diverse ways, the rate of progress will be limited until ignition and ignition with modest gain are attained. The event-based roadmap provided in this chapter uses these two events (ignition and modest gain) as early milestones that could trigger the creation of a robust IFE program.

INTRODUCTION

The development of any science- or technology-based roadmap requires that guidelines and criteria be established so that options are evaluated on a common and consistent basis. The committee believes that the guidelines detailed in the DOE Technology Readiness Assessment Guide[1] are useful and appropriate for the development of an IFE roadmap, so the committee uses them here. Figure 4.1 shows the integration between technology development and project management. As can be seen from the chart, a conceptual design is created at the CD-0 point (yellow box) in a project.

As suggested in DOE G 413.3-4A, a useful and recommended approach to assure that the various technical components are at a stage of technical maturity suitable to initiate the next phase in the program is used—the concept of "technology readiness levels" (TRLs). The TRLs of the overall system as well as its components must be advanced and evaluated over time. Table 4.1 lists the nine TRLs discussed in the Guide, which contains more detailed descriptions of the TRLs.

In keeping with the Guide, the committee has assumed that all necessary technology options and their components must have met the criteria of TRL 6 for DOE to initiate the conceptual design for an IFE DEMO. Development activities and test facilities, including major test facilities such as integrated test facilities (IREs) and an FTF, as defined in Box 4.1, will help to advance the TRLs of components necessary for DEMO. However, the components of an IRE and an FTF must also

---

[1] U.S. Department of Energy, *Technology Readiness Assessment Guide*, DOE G 413.3-4A, October 12, 2009. Available at http://tinyurl.com/84qk6qw.

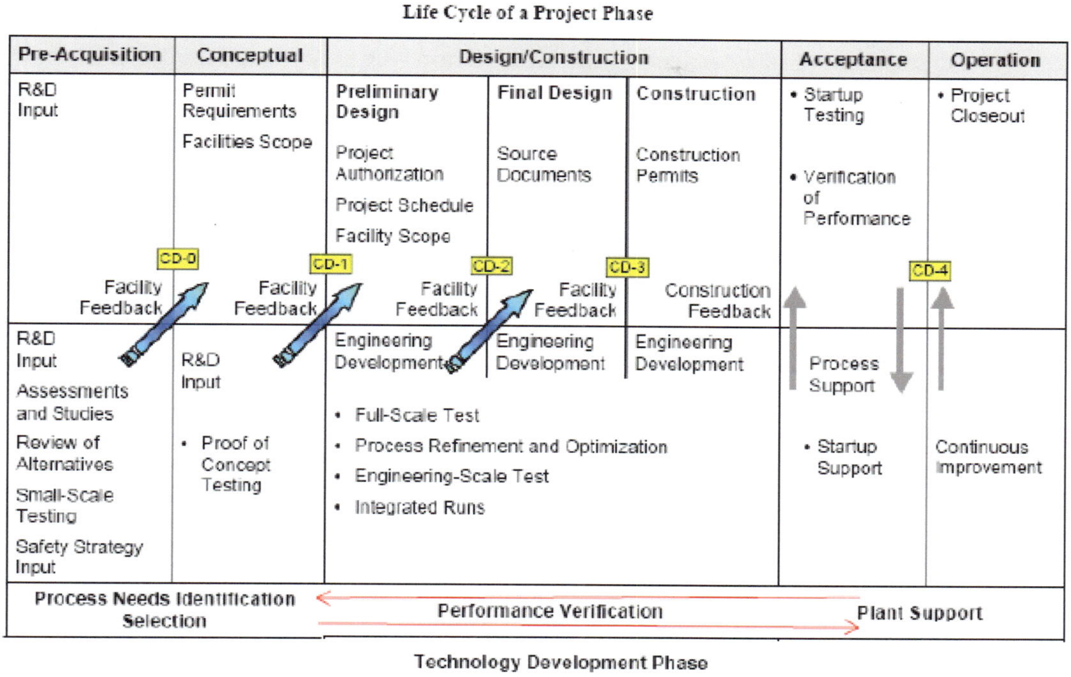

FIGURE 4.1 Process and performance requirements to support plant startup, commissioning, and operations. SOURCE: DOE Technology Readiness Assessment Guide.

TABLE 4.1 Technology Readiness Levels

Basic Technology Research
   TRL 1: Basic principles observed and reported
   TRL 2: Technology concept/application formulated
Research to Prove Feasibility
   TRL 3: Proof of concept
Technology Development
   TRL 4: Validation in laboratory environment
   TRL 5: Integrated component validation in laboratory
Technology Demonstration
   TRL 6: Engineering/pilot scale validation
System Commissioning
   TRL 7: Prototypical system demonstration
   TRL 8: System qualified through test and demonstration
System Operations
   TRL 9: Full range of actual system operations

have reached certain TRLs in order for those facilities to be built. The TRLs for each IFE option are summarized in the later section "TRLs for Inertial Fusion Energy."

TECHNOLOGY APPLICATIONS

There are many possible combinations of drivers, targets and chambers that could be considered as technology applications (TAs). As mentioned above, the committee has defined three TAs based on the three main driver approaches: lasers, pulsed power, and heavy ions. These three TAs cover the main options for targets, drivers, and chambers, and simplify the task of planning an event-based roadmap. For example, the heavy-ion fusion plan would require the research needed to select between radio-frequency and induction accelerators and an approach to target design. Similarly, the laser TA would have to consider the research needed to decide between diode pumped solid state laser (DPSSL) and KrF laser drivers and between direct and indirect drive. The focus is on the research needed to make decisions and to optimize progress rather than to sustain a particular TA as long as possible. Thus, eventually, either a single TA would be taken to the DEMO stage or no TA would be judged to be both technically feasible and economically viable.

For each TA, the driver is the most expensive component in the power plant. In all three cases, the driver will consist of a large number of modules. As discussed in Chapter 2, good progress has been made in developing the repetitively pulsed systems required for fusion energy. Nevertheless, there remain substantial challenges in developing systems that would have the reliability, maintainability, and availability to provide between 3×10^6 and 4×10^8 shots per year, depending on the driver. As concluded in Chapter 2, it will be necessary to build and demonstrate each multikilojoule module early in the program.

> **Recommendation 4-1: When a technical approach is chosen, high priority should be given to the design and construction of a driver module and to demonstrating that the individual driver module meets its specifications so that when aggregated into a complete system, the appropriate gain can be achieved.**

Institutional competition has been important in driving innovation in IFE, as it has been in many fields. At this point in time, however, the IFE community would benefit from greater cooperation and integration. A recent white paper developed by the IFE community reached the same conclusion.[2] Without a coordinated

---

[2] M. Hockaday, N. Alexander, S. Batha, M. Cuneo, M. Dunne, G. Logan, D. Meyerhofer, A. Nikroo, S. Obenshain, D. Rej, and J. Sethian, "White Paper Compilation on Inertial Fusion Energy (IFE) Development," March 30, 2011.

approach to IFE, it will be difficult for the nation to make informed decisions using reliable cost estimates and confidence levels.

Within heavy-ion fusion, there is almost no difference in the research programs needed for direct drive and indirect drive in the near term. The beam requirements for the two options are sufficiently similar that it is not necessary to split the approaches into two TAs. At some point in the future, however, there is a key choice to be made between these two options. The existence of a Virtual National Laboratory for HIF has facilitated thinking about the program as a single TA. The multiple institutions involved in heavy-ion fusion research work together closely, and no institution is threatened when a major decision is made. There are enough internal advocates of various approaches to maintain innovation, but DOE should monitor this to assure that innovation remains active.

In contrast, the competition between the various approaches for laser-driven, heavy-ion-driven, and pulsed-power-driven fusion is led by institutions, each of which advocates a different approach. The IFE effort would benefit greatly from a joint plan together with an approach to program governance that can make difficult decisions but is able to retain the strengths of all the institutions. Virtual laboratories could well serve the decision analysis required to advance IFE research. Two examples of such virtual laboratories are given in Box 4.2.

A virtual laboratory can facilitate difficult decisions involving programmatic direction. For example, LLNL began building a small recirculating induction accelerator while LBNL was working on the more standard linear induction accelerator. It became apparent that one could not sensibly carry out both approaches with

---

BOX 4.2
Virtual Laboratories

The Virtual Laboratory for Technology (VLT) was created in 1999 by DOE's Office of Fusion Energy Sciences (OFES) to coordinate and represent all magnetic fusion technology activities funded by OFES. It is an on-going national activity. The scope of activities includes or has included plasma heating and fueling technologies, magnet systems, plasma facing components, fusion nuclear technologies including tritium-breeding blankets, fusion safety analysis, research on advanced materials, and fusion systems studies and analysis. A wide variety of national laboratories, universities, and industry are or have been members of the VLT.

The Heavy-Ion Fusion Virtual Laboratory (HIF-VL) was created in the mid-1990s. It was created with a formal agreement among LLNL, LBNL, and the Princeton Plasma Physics Laboratory (PPPL). The director of the HIF-VL has been from LBNL since LBNL has the largest program of the three laboratories. The two deputy directors are from LLNL and PPPL. Their meetings and seminars are frequent and are handled by teleconference. LLNL representatives have offices at LBNL, which also facilitates communication.

realistic budgets, so a choice between the two was necessary. The laboratories had the requisite expertise to make a technical decision, but DOE did not, so the HIF-VL took the lead and a decision was reached. An analogous situation for lasers would be a choice between the KrF and DPSSL laser, for example. If there is not enough funding to pursue both options, a choice will have to be made. A virtual laboratory can help keep the discussion of technical decisions at the technical level and avoids nontechnical considerations that can prevent optimal decisions from being reached.

Conclusion 4-1: The focus of any formal inertial fusion energy program would be best served if the program were organized according to the three Technical Applications (TAs): laser systems, heavy-ion systems, and pulsed power systems.

To accomplish this organization, several actions are recommended.

Recommendation 4-2: The national inertial fusion energy program should be organized according to three technical applications: laser systems, heavy-ion systems, and pulsed power systems.

Recommendation 4-3: The Department of Energy should consider the establishment of virtual laboratories for each technical application with sufficient internal expertise for the various approaches to advance technically and maintain innovation.

EVENT-BASED ROADMAPS

In Chapters 2 and 3 the committee discussed the status of the driver options, including the targets and various fusion technologies, for each approach under consideration for IFE. In doing so, it came to several general conclusions that help govern the development of a composite roadmap.

In Chapter 2 the committee came to some general conclusions:

- There are a number of technical approaches, each involving a different combination of driver, target, and chamber, that show promise for leading to a viable IFE power plant. These approaches involve three kinds of targets: indirect drive, direct drive, and magnetized target. In addition, the chamber may have a solid or a thick-liquid first wall that faces the fusion fuel explosion.
- Substantial progress has been made in the last 10 years in advancing most of the elements of these approaches, despite erratic funding for some

programs. Nonetheless, a substantial amount of R&D will be required to show that any particular combination of driver, target, and chamber would meet the requirements for a demonstration power plant.
- In all cases, the drivers may build on decades of research in their area. In all technical approaches there is the need to build a reactor-scale driver module for use in an FTF.

Similarly, the committee stated three general conclusions in Chapter 3. First, it said that technology issues—e.g., chamber materials damage, target fabrication, and injection, etc.—can have major impacts on the basic feasibility and attractiveness of IFE and thus on the direction of IFE development. Next, it concluded that at this time, there appear to be no insurmountable IFE fusion technology barriers to the realization of the components of an IFE system, although knowledge gaps and large performance uncertainties remain, including for the performance of the system as a whole. And finally it determined that significant IFE technology research and engineering efforts are required to identify and develop solutions for critical technology issues and systems, such as: targets and target systems; reaction chambers (first wall/blanket/shield); materials development; tritium production, recovery and management systems; environment and safety protection systems; and economics analysis.

Thus, each of the three TAs, as the committee has defined it above, has to achieve certain significant milestones, or events (e.g., ignition) before it can logically move on to the next step. What is needed is a scientific understanding of gain and target design for robust operation—not just gain. For example, (1) ignition, (2) reactor-scale gain, (3) reactor-scale gain with potential cost-effective targets, and (4) reactor-scale gain at high repetition rate are examples of milestone events that must be satisfactorily achieved before going on to the next step:

Interval 1 Interval 2 Interval 3 Interval n Interval $n + 1$
----------(event)----------(event)----------(event)--//-------(event)---------DEMO

For each interval one needs to consider the following:

- Significant development(s) required,
- Potential scientific and technological roadblocks,
- Required facilities, existing or new (if a new facility is needed, one must indicate when it needs to be started (CD-0) (see Figure 4-1),
- Synergies with the magnetic fusion energy (MFE) program, and
- Estimated costs to accomplish activities in each interval.

The significant events that are listed above are target- and driver-centric because ignition has not yet been achieved in ICF, but target and driver concerns are not the only issues facing inertial fusion. Chambers (materials) that survive and that are economical must also be found. For laser-driven systems, optics that survive and retain their optical quality for a long time in an adverse environment must exist. The drivers not only must achieve the desired repetition rate, but also must achieve durability and reliability objectives. The cost of the drivers must be acceptable. A given TA could march relatively easily through a given set of significant science-based events but still fail as a power plant because of technology and economic considerations.

Each TA will require years of research and development before a DEMO can be designed in any detail. No TA has yet demonstrated fusion gain, reactor-level driver energy at repetition rate, or chamber life.[3]

In summary, the following criteria (events) must all be satisfied before committing to a DEMO.

1. First and foremost, ignition must be demonstrated. Absent ignition, any IFE program will be severely limited in scope.
2. Modest (or adequate) gain must be demonstrated to a level relevant to the TA[4] in question to ensure that the TA has a feasible technical approach to achieving high gain.
3. Target gain must be demonstrated at the relevant high level, which varies with each TA, depending on the driver efficiency. One guideline, based on basic power balance considerations, is that the product of driver efficiency times the gain should be greater than or equal to 10. Obviously, having a margin on this requirement would be an advantage. Table 4.2 contains estimates of driver efficiency—supported by component and subsystem tests—and goals for reactor-scale gain that are supported by theoretical modeling and computer simulations for the various approaches.
4. Driver life at energies corresponding to the reactor-scale gain level must be demonstrated to $>10^7$ pulses (except pulsed power, which must be demonstrated to $>10^6$ pulses) and must extend in predictable ways to 100 times greater than 10^7 (or 10^6) pulses before commitment to an FTF or DEMO.
5. Target fabrication for each TA has to be automated at a level related to the target consumption in the FTF and must extend predictably to the DEMO consumption level at costs consistent with a competitive cost of electricity.

---

[3] Appendix J indicates the steps required for each TA to reach the starting point of the DEMO conceptual design. The specific steps are meant to be illustrative of the conditional requirements that DOE should set down in its planning process—requirements that should be regularly updated based on scientific and technological progress.

[4] The relevant gain varies with each technical approach and depends on the driver's efficiency. See Table 4.2.

TABLE 4.2 Driver Efficiencies and Minimum Gains That Will Be Required to Demonstrate the Viability of Reactors Based on Various Driver Technologies

Technology Approach	Estimated Driver Efficiency η_D (%)	Reactor-Scale Gain $\eta_D \times G > 10$
Solid-state lasers	16	>60
KrF lasers	~7	>140
Heavy-ion beams	25-45	20-40
Pulsed power	20-50	20-50

NOTE: The numbers in this table are only illustrative and are not meant to be definitive.

6. Chamber design, including neutron shielding, tritium breeding, and materials survival, has to be sufficiently developed to generate a high probability of successful operation for multiple years. It is not possible to fully test the chamber design under fusion conditions short of executing an FTF or a DEMO. One of the strongest reasons for an FTF to precede a DEMO is to validate the chamber design.

The most appropriate ordering of the milestones in a roadmap will differ for different driver/target combinations.

Conclusion 4-2: Despite the significant advances in inertial confinement fusion, many of the technologies needed for an integrated inertial fusion energy system are still at an early stage of technological maturity. For all approaches to inertial fusion energy examined by the committee (diode-pumped lasers, krypton fluoride lasers, heavy-ion accelerators, pulsed power; indirect drive and direct drive), there remain critical scientific and engineering challenges associated with establishing the technical basis for an inertial fusion energy demonstration plant. It would be premature at the present time to choose a particular driver approach as the preferred option for an inertial fusion energy demonstration plant.

It is clear that reactor-scale gain must be uniquely defined for each TA since the understanding of gain involves laser-plasma interaction physics, hohlraum physics (for indirect drive only), ablation physics, instabilities and mix, symmetry control, equations of state, real-world fabrication and alignment tolerances, and temperature control.

Conclusion 4-3: Owing to the technical complexity, the specific definitions of modest (or adequate) and high gain should be determined independently for each technology application.

COMPOSITE ROADMAP AND DECISION ANALYSIS FOR THE PRE-IGNITION STAGE

Given that there are many variables and options to consider before being able to proceed with the conceptual design for a DEMO plant, the committee believes it would be most useful to focus on the earliest stage—namely, pre-ignition—by adding a decision-tree analysis to only this first phase of the roadmap.[5] The immediate future is the most clear, and it is also the most critical time for IFE as the National Nuclear Security Administration (NNSA) program strives to demonstrate ignition. Accordingly, the committee's analysis was based on the effort at the National Ignition Facility (NIF) in 2011-2012 to achieve ignition under the National Ignition Campaign (NIC). Pre-ignition contingency planning was considered in more detail, but the details have not been included here because events and NNSA's path forward have changed the basis for such a plan; however, the committee believes that event-based, decision-tree analysis (contingency planning) is important for a complex, multifaceted program such as IFE.[6]

ICF research has been driven by NNSA for stockpile stewardship requirements. The decision to build the NIF, which is designed to operate in single-shot mode and is not currently equipped to serve as a test facility for repetition-rated operation or engineering tests for IFE, was based upon those requirements. NIF conducted the NIC with the end objective being ignition by the end of FY2012, Having reached the end of the NIC campaign on September 30, 2012, without achieving ignition, NNSA decided to revise the operational program for NIF.[7]

Given the substantial investment already made in the NIF, from the NNSA perspective, laser indirect-drive is the preferred approach for stockpile stewardship if ignition with sufficient yield for the desired experiments can be achieved. When one considers the application of ICF for the production of practical electric power in the

---

[5] C.B. Chapman and S. Ward, 2003, *Project Risk Management: Processes, Techniques, and Insights*, 2nd ed., Hoboken, N.J.: Wiley.

[6] To assist in its thinking about pre-ignition contingency planning across TAs, the committee prepared several detailed hypothetical examples. The common elements are included in the text.

[7] On December 8, 2012, NNSA released its report to Congress, "NNSA's Path Forward to Achieving Ignition in the Inertial Confinement Fusion Program" (hereinafter referred to as "NNSA Path Forward 2012 Report to Congress"). This report represents the views of the NNSA and was prepared principally by program representatives from the ICF laboratories and other principal contractors through participation in various working groups. The NNSA report proposes a time-based (3-year) plan. The report describes the path forward for NIF as requiring a transition from the NIC to a facility with greater focus on the broader scientific applications of NIF and a priority on key questions regarding stockpile stewardship. For IFE pre-ignition efforts, the approach advocated by the NRC Committee on the Prospects for Inertial Confinement Fusion Energy Systems is event-based (as opposed to time-based) and thus might not be limited to 3 years, and might include TAs not considered in the NNSA's 3-year plan.

context of organizing research through an IFE program, other equally critical steps become apparent—namely, achievement of reactor-scale gain, reactor-scale gain with a cost-effective target, and reactor-scale gain with the required repetition rate.

Conclusion 4-4: The schedule for each technical application is driven by the time required to demonstrate certain milestones, while the composite inertial fusion energy roadmap is focused on a single DEMO. Implementation of the roadmapping process can provide a very useful tool to determine the appropriate course of action.

Decisions will need to be made about the continuation of individual TAs in the absence of significant progress. The dilemma, then, is the balance between the continuation of the three major TAs in contrast to an early down-select process that would define the TA for application to DEMO. The roadmapping process can be very useful in determining the appropriate course of action.

It must be recognized that roadmapping, as discussed here, is a snapshot in time and needs to be revisited on a periodic basis or when a single significant event occurs. The process is meant to be continually informed by these periodic snapshots of where the science and technology stand relative to the goal of achieving CD-0 (see Figure 4.1) for DEMO. Using TRLs to assess the technical maturity of the various components will be necessary to inform the roadmapping process.

Recommendation 4-4: The Department of Energy should use a milestone-based roadmap approach, based on technology readiness levels (TRLs), to assist in planning the recommended national IFE program leading to a demonstration plant. The plans should be updated regularly to reassess each potential approach and set priorities based on the level of progress. Suitable milestones for each driver-target pair considered might include, at a minimum, the following technical goals:

1. **Ignition,**
2. **Reproducible modest gain,**
3. **Reactor-scale gain,**
4. **Reactor-scale gain with a cost-effective target, and**
5. **Reactor-scale gain with the required repetition rate.**

Coupling the physics of the driver-target to the system that can extract the energy is a serious engineering challenge. The ability to inject and ignite a target, capture the energy released, clear the ignition chamber, and then repeat the whole process multiple times per second is a major technical issue for IFE. Coupled physics-engineering tests will be needed to develop solutions.

It is assumed in the following discussion that NIF, which was designed for a 30-year lifetime, continues operation after 2012. Until the results of the current ignition campaign have been analyzed, it will be difficult to decide the extent to which resources and beam time should be given to the various experiments and upgrades that should be considered for NIF. For that reason, the committee recommends below that a science advisory committee focused on IFE be formed to advise decision makers on detailed allocations of resources and beam time for NIF as well as to develop the post-ignition roadmap.

Recommendation 4-5: Future inertial fusion energy-related experiments on the National Ignition Facility should be reviewed by an Inertial Fusion Energy Scientific Advisory Committee (ISAC) as one of its first tasks, and it should be established in consultation with the Department of Energy and be comprised of technical experts for all options being considered, including experts who can serve as referees.

Two philosophies regarding the development of IFE were evident in the literature and in the presentations made to the committee. One approach emphasizes looking for existing technology, grounded in existing knowledge, to engineer fusion components, unless or until a roadblock appears, at which point science and technology research are applied to overcome the obstacle. This approach may speed up the DEMO process by identifying solutions to known problems but may not result in an optimal design.

The second philosophy, which is more systematic, is aimed at understanding each phenomenon by means of scientific and technological research before moving on to the next step. This approach, while possibly slower in producing a DEMO, may allow optimization of a DEMO.

Historically, the two philosophies have found homes in different approaches to developing IFE. Although all approaches contain elements of both, the first is exemplified by the laser inertial fusion energy (LIFE) program[8] and the second by the High Average Power Laser (HAPL) program.[9] A priori, there is no correct

---

[8] The LIFE program is an integrated engineering study of an IFE plant facility (DEMO) that combines the best of what is available in technology with input from customers (utilities), from engineering capability (large engineering companies), and from experiments under way to achieve ignition on targets (government). The key ingredient is to design to meet user needs supported by the available technology, with R&D aimed at risk mitigation undertaken by government. The LIFE study has been supported by LLNL laboratory directed R&D (LDRD) funds at $10 million per year over the past 4 years.

[9] The HAPL program was an integrated program mandated by Congress from FY2001 to 2009 to develop the science and technologies for fusion energy using laser direct drive. It was managed by the Naval Research Laboratory (NRL) and involved 7 government laboratories, 8 universities, and

balance between these different philosophies. Balance is achieved by the exercise of subjective judgment that may vary depending on the development stage of IFE, the personal experience of the researchers, and even the political philosophy of government administrations. It is important that the competition between these two approaches not interfere with the best use of the NIF facility for IFE development.

The pre-ignition roadmap described in this report is meant to be an example of the kind of contingency planning that the committee believes DOE should undertake across TAs, with the advice and review of the Inertial Fusion Energy Scientific Advisory Committee, as recommended above. If at any time ignition is reached for any TA, the roadmap would shift from pre-ignition to post-ignition.

Ignition hopes and efforts have been focused primarily on indirect drive on the NIF. Even though ignition was not reached by the end of FY2012, it will be a number of years (approximately 2017) before alternative approaches, such as direct drive in the form of a polar direct drive configuration, could be tested on the NIF. This should give ample time to understand why the model predictions of indirect drive's performance were invalid and to try new approaches with indirect drive using the current NIF configuration, should new understanding warrant them.

With ignition not having been achieved with laser-indirect drive, a commitment would be warranted to build the optics and other components for a polar direct drive option on the NIF, recognizing that the completed system could not be operational for 4 or more years.[10] As a first step, it would be appropriate to measure the extent of laser-plasma instabilities and experiment with beam smoothing, both of which are precursor activities that can be done before installing polar direct drive (2017, at the earliest). Deciding on the balance of these experiments and experiments appropriate for understanding the failure of indirect drive to achieve ignition by the end of the NIC could be informed by the ISAC, identified in Recommendation 4-5. Note that even if ignition is reached with indirect drive before 2017, a decision to build the polar drive option would be warranted to explore opportunities for higher gain. And, modification of NIF to polar direct drive would not foreclose future experiments with indirect drive, although some setup time would be required to switch configurations.

If polar direct drive on NIF should show promise that direct drive might well reach ignition, construction of a spherical direct drive system for the NIF would be

---

17 companies, with annual budgets around $15 million. Through it, sufficient progress was made in developing repetitively pulsed DPSSL and KrF lasers to give confidence that both concepts were worth considering for IFE. Progress was also made on target launching and tracking, final mirror optics, frozen tritium behavior, first wall materials issues, magnetic diversion to protect the first wall, and systems studies. See http://aries.ucsd.edu/HAPL.

[10] LLNL, 2012, *Polar Drive Ignition Campaign Conceptual Design*, LLNL TR-553311, submitted to NNSA in April and revised and submitted to NNSA by LLE in September 2012.

the next step. Again, a spherical direct drive system would not rule out continuing tests with indirect drive by using approximately two-thirds of the beams.

If both the laser-indirect and laser-direct drive approaches continue to experience difficulty reaching ignition over the next 5 or so years, then it would be justified to put more resources to the MagLIF and HIF approaches. Depending on the reasons for the failure of the laser-based approach—e.g., laser plasma instabilities—it might also be appropriate to consider alternate laser driver approaches. DOE support for reactor design studies of ideas using these drivers is important, including participation by groups that are not advocates. Viable reactor designs would be required before there is a substantial ramping up of such approaches. These design studies should help guide the related decisions.

> **Recommendation 4-6: Although ignition was not achieved at the National Ignition Facility by the end of FY2012 as planned, efforts to achieve ignition with indirect drive should not cease. Contingent on the availability of funds and Department of Energy priorities, these efforts should continue at least until new configurations (e.g., polar direct drive) can be tested on the National Ignition Facility, which would require at least 4 years of development. However, under this scenario, a commitment should be made to undertake pretesting of polar direct drive on the National Ignition Facility and, if the pretests are successful, prepare NIF to test polar direct drive.**

Even if ignition should be reached with indirect drive before polar direct drive becomes operational, the funding for direct drive will still have been well spent, for it is desirable to test polar direct drive in the hope of getting a higher gain (with the same drive energy) than may be possible with indirect drive. (A technical discussion of direct and indirect drive is given in Chapter 2.)

As discussed in Chapter 2, the energy required to achieve ignition in laser-based indirect and direct drive approaches favors direct drive. Moreover, for a fixed laser energy, the calculated gain is higher for direct drive. Nevertheless, there are important uncertainties in laser-plasma physics and implosion dynamics that must be addressed for fusion-scale targets, particularly for shock ignition. The NIF is currently a unique tool for addressing these issues, some of which could be addressed with NIF in its present configuration. Others may require modifications such as improvements in beam smoothness, or ultimately even a different illumination geometry.

> **Conclusion 4-5: Each target design and each driver approach has potential advantages and uncertainties to the extent that "the best driver approach" remains an open question.**

Recommendation 4-7: The achievement of ignition with laser-indirect drive at the National Ignition Facility should not preclude experiments to test the feasibility of laser-direct drive. Direct drive experiments should also be carried out because of their potential for achieving higher gain and/or other technological advantages.

Conclusion 4-6: It is essential for the IFE program to develop reliable models and improve the physics understanding of the phenomena underlying experimental tests of the target physics. Knowledge gained through experimental tests should be used to validate and improve the models, so that there can be reasonable confidence that the predictions are not restricted to the parameter space explored in the experimental tests. Models will be important for optimizing designs from both a technological and economic perspective.

Conclusion 4-7: Achieving higher gains has the potential to provide improved technical margins and potential economic advantages for the system as a whole. If calculations are confirmed, fewer targets would be needed to produce a given amount of power, or the driver repetition rate or driver energy could be reduced, thereby reducing costs.

TRLs FOR INERTIAL FUSION ENERGY

An important question is which facilities will need to be built to successfully reach the goals of the IFE program. Table 4.3 is based on the data provided in the discussions in Chapters 2 and 3 on the TAs respecting what has been done and what is under way in IFE, as well as what the magnetic fusion energy program provides and what needs to be done to reach the conceptual design stage of DEMO and commercial deployment of IFE. In addition to a number of smaller test facilities (i.e., IREs), it assumes that there will be an additional two major facilities: (1) a Fusion Test Facility (FTF), a staged facility with repetitively targeted deuterium-tritium (DT), high-gain capsules that would bring all aspects of the technology of IFE up to TRL 6 using a prototypical driver that would be determined by the IFE program, and (2) the end point of the IFE development program, DEMO, which would complete the TRL process.

As shown in Table 4.3, NIF and FTF are absolutely critical to move the TAs and their technological components from TRLs of 4 or less to 6 for the CD-0 DEMO decision process. Note also that it has been assumed in Table 4.3 that certain technologies (e.g., materials, handling) will be developed, at least in part, using existing MFE facilities, as described in Chapter 3.

TABLE 4.3 Facilities and Efforts Required to Advance Fusion Energy Technologies to Various Technology Readiness Levels (TRLs)

Area	TRL								
	1	2	3	4	5	6	7	8	9
Target physics	Weapons, OMEGA, etc.		NIF	FTF			DEMO		
Target manufacture	GA work, HAPL		NIF	ATFF/FTF			DEMO		
Drivers[a]	Depends on system			FTF			DEMO		
Control[b]	HAPL NIF			FTF			DEMO		
Diagnostics	OMEGA, etc.		NIF		FTF		DEMO		
Materials[c]	MFE			IFMIF	FTF		DEMO		
Tritium breed	MFE, lab tests liquids			ITER	FTF		DEMO		
Tritium system	JET, TFTR, TSTA			ITER		FTF	DEMO		
Power handling				ITER, FTF			DEMO		
Remote handling	JET					ITER, FTF	DEMO		
Reliability	FTF						DEMO		
Availability	FTF						DEMO		
Safety	NIF				ITER, FTF		DEMO		
Waste handling	TFTR, JET, fission facilities, ITER, FTF						FTF	DEMO	

NOTE: NIF, National Ignition Facility; FTF, Fusion Test Facility; DEMO, Demonstration Power Plant; HAPL, High Average Power Laser program; GA, General Atomics; ATFF, Automated Target Fabrication Facility; MFE, magnetic fusion energy; IFMIF, International Fusion Materials Irradiation Facility; ITER, International Thermonuclear Experimental Reactor; JET, Joint European Tokamak; TFTR, Tokamak Fusion Test Reactor; and TSTA, Tritium System Test Assembly.

[a] The various drivers are at different TRLs in FY2012. For example: NIF single-shot laser, TRL 9; Repetition rate of IFE solid state lasers, TRL 4; Heavy-ion beams: TRL 3 to TRL 6, if existing but different accelerators are taken into account; Pulsed power, TRL 5.

[b] Present targets are fixed. Repetitive targeting of DT targets on the fly will have to wait for FTF.

[c] The answer depends on which type of first wall is considered—thick liquid wall, thin liquid wall, or solid wall.

Conclusion 4-8: There are several technology development areas in which there is overlap and/or synergy between magnetic fusion energy (MFE) and inertial fusion energy (IFE).

Recommendation 4-8: The overlap and synergies that exist between MFE and IFE technology development areas should be exploited. The Department of Energy should assure that the research program plans for IFE and MFE are coordinated and that the research results are fully shared between the two programs.

COST AND FUNDING CONSIDERATIONS

The further one looks into the future, the more difficult it is to estimate what the appropriate budget levels should be. Not only are there variables in the budgeting process, there are also uncertainties as to the probability of achieving the research objectives and milestones identified in this report as well as to the length of time needed to achieve them. What makes planning particularly difficult is the fact that three competitive approaches exist, and, ultimately, only one can be selected as the TA for the DEMO.

Research in ICF is currently funded largely by NNSA and involves the weapons laboratories (LLNL, LANL, SNL), NRL, and a number of university-managed laboratories, most notably the Laboratory for Laser Energetics (LLE) at the University of Rochester and LBNL. The major experimental facilities are the laser facilities NIF (at LLNL), OMEGA (at LLE) and NIKE (at NRL), and the pulsed power system Z at SNL. The weapons laboratories and a number of universities house smaller facilities. A Virtual National Laboratory for Heavy Ion Fusion Science consisting of LBNL, LLNL, and the Princeton Plasma Physics Laboratory undertakes the heavy-ion fusion program; its present work is focused on high-energy-density physics and heavy ion fusion science and is funded by DOE's Office of Fusion Energy Sciences. The magnetized target fusion approach is studied by LANL and the Air Force Research Laboratory.[11]

Previous funding sources for IFE R&D have been diverse and have included Laboratory Directed Research and Development (LDRD) funds at the NNSA laboratories—for example, Laser Inertial Fusion Energy (LIFE) and pulsed power approaches—direct funding through the Office of Fusion Energy Sciences (e.g., heavy ion fusion, fast ignition, and magnetized target fusion), and congressionally-mandated funding. Beginning in FY1999, Congress directed the initiation of the HAPL program, to be managed by NNSA. The HAPL program was an integrated program to develop the science and technology for fusion energy using laser direct drive. Initially focused on the development of solid-state and KrF laser drivers, HAPL then expanded to address all of the key components of an IFE system, including target fabrication, target injection and engagement, chamber technologies and final optics, and tritium processing.

Currently, by far the largest support for ICF comes under the NNSA Stockpile Stewardship program, which supports LLNL's activities (including NIF), the program on the OMEGA laser at the University of Rochester, the use of KrF lasers at NRL, and Sandia's pulsed-power efforts on the Z facility. Within this NNSA program, the main focus was the NIC at NIF. The NIC carried out a 200-shot program on the NIF managed by LLNL. The sequence of shots was focused on a stepwise

---

[11] See Chapter 2 for more discussion on the activities at these institutions.

progression in driver beam power and intensity, including shock timing, optical focus, mix, and target-hohlraum geometries. The schedule called for the 200-shot NIC program to culminate in ignition by the end of FY 2012. As discussed in Box 1.2 and Appendix I, ignition was not achieved by the end of the NIC.

Conclusion 4-9: While there have been diverse past and ongoing research efforts sponsored by various agencies and funding mechanisms that are relevant to IFE, at the present time there is no nationally coordinated research and development program in the United States aimed at the development of inertial fusion energy that incorporates the spectrum of driver approaches (diode-pumped lasers, heavy ions, krypton fluoride (KrF) lasers, pulsed power, or other concepts), the spectrum of target designs, or any of the unique technologies needed to extract energy from any of the variety of driver and target options.

Conclusion 4-10: Funding for inertial confinement fusion is largely motivated by the U.S. nuclear weapons program, due to its relevance to stewardship of the nuclear stockpile. The National Nuclear Security Administration (NNSA) does not have an energy mission and—in the event that ignition is achieved—the NNSA and inertial fusion energy (IFE) research efforts will continue to diverge as technologies relevant to IFE (e.g., high-repetition-rate driver modules, chamber materials, mass-producible targets) begin to receive a higher priority in the IFE program.

The largest technology component of the NNSA stockpile stewardship budget deals with target physics. Based on information provided to the committee, this support appears to be around $260 million per year.[12] At this stage the objectives for target physics of the NNSA's ICF program are relevant to the inertial fusion energy program. While NNSA will continue to have an interest in target physics research after ignition is achieved, it will become less critical to meeting national security objectives, and there will be less overlap with the needs of IFE. For example, an IFE target may need to have a higher yield than NNSA would normally be interested in, and NNSA might not be interested generally in certain approaches. Accordingly, NNSA is unlikely to undertake technology research that is relevant only to fusion energy (e.g., chambers).

Conclusion 4-11: If a coordinated national program in inertial fusion energy is established, one of the first orders of business will be to resolve responsibility

---

[12] Jeffrey Quintenz, "Status of the National Ignition Campaign and Plans Post-FY 2012," Presentation to the committee on February 22, 2012.

and budgeting for target physics work, understanding that the needs for the inertial fusion energy program diverges from those for stockpile stewardship.

While existing NNSA facilities (NIF, Z, OMEGA) are critical to the IFE effort, this report notes that, in order to reach the CD-0 stage for a DEMO plant, other facilities will need to be built, and these, in turn, must also go through the various project phases and decisions (CD-0 through CD-4). The largest and most important precursor facility for IFE is an FTF. As evident from the preceding discussion, the design of the FTF should begin at a propitious time in order to start tritium operations of the FTF in a timely manner and to have data for input to the DEMO project decision process.

Conclusion 4-12: Existing facilities (NIF, Z, OMEGA, NDCX-II, HCX, NIKE, and Electra) will play critical roles in advancing the technical applications and their technological components from technical readiness levels (TRLs) of 4 or less to TRL 6 for the CD-0 demonstration plant (DEMO) decision process. In addition, to have a successful national IFE program, adequate funds are required to implement one or more integrated research experiments, at least one Fusion Test Facility, and the upfront costs for the DEMO design.

Table 4.4, based on the inputs to Chapters 2 and 3 and the above considerations, provides a rough outline of the near-term programmatic funding requirements if an IFE program were to proceed in a two-step ramping process with annual budgets of at least $50 million after ignition is attained and some $90 million-$150 million after ignition plus modest gain has been demonstrated. Table 4.5 contains an order-of-magnitude estimate of future minimum capital cost requirements for an IFE program.

It is difficult to provide an overall programmatic cost estimate since there are several significant uncertainties that have to be resolved, such as the length of time required to reach the decision on DEMO, the ability to successfully complete milestones in a timely fashion, the extent to which each TA will be pursued, the number of IREs that will be required, and whether more than one FTF will be built. In 2003, the Fusion Energy Sciences Advisory Committee (FESAC) made a combined magnetic fusion energy and inertial fusion energy programmatic cost estimate.[13] Based upon that report and the LIFE point design forecast,[14] the committee's order-of-magnitude estimates for facility capital costs, subject to the DOE G 413.3-4 process, are provided in Table 4.5.

---

[13] FESAC, Fusion Development Panel, 2003, *A Plan for the Development of Fusion Energy*, March.

[14] T. Anklam, M. Dunne, W.R. Meier, S. Powers, A.J. Simon, LIFE: The case for early commercialization of fusion energy, *Fusion Science and Technology*, 60: 66-71.

TABLE 4.4 Estimated Near-Term Inertial Fusion Energy Roadmap Development Cost Forecast, After Ignition[a]

	Annual Budget (millions of 2012$)	
Technology Application	Post-Ignition	Post-Ignition/Modest Gain
DPSSL/KrF lasers[b]	20-30[c]	40-60[d,e]
HIF	~10	20-30
Pulsed power	~10	10-20
Technology development	10-20	20-40
Totals	50-70	90-150

[a] The values given are capital and development costs and do not include operating costs.
[b] Michael Dunne, LLNL, Presentation to the committee on February 22, 2012, and subsequent communications.
[c] Ibid.
[d] Ibid.
[e] This is the estimated annual cost over 3 years to build and commission the single beam line laser source for LIFE.

TABLE 4.5 Estimated Inertial Fusion Energy Roadmap Facility Capital Cost Forecast (millions of dollars)[a,b,c]

Facility	Cost
NIF upgrade (polar drive)	50-60[d,e]
NIF upgrade (spherical drive)[f]	Unknown[g]
IRE	300-775
FTF	3,100-4,750
DEMO	6,250-9,500

[a] All values include a 25 percent contingency.
[b] All numbers have been escalated from 2002$ to 2012$ using the Office of Management and Budget's GDP (Chained) Price Index (estimate for 2012), except for the NIF upgrade (polar drive), which is given in as-spent dollars.
[c] All costs are capital costs and are subject to the DOE G 413.3-4 process.
[d] Cost for the procurement of unique hardware, optics, and controls systems.
[e] LLNL, 2012, "Polar Drive Ignition Campaign Conceptual Design," LLNL TR-553311, submitted to NNSA in April 2012 by LLNL and revised and submitted to NNSA by LLE in September 2012.
[f] If needed to obtain high gain. Some of this cost might be covered as part of the stockpile stewardship program if sufficient gain is not obtained with indirect drive.
[g] The committee is unaware of any detailed cost estimate for this upgrade. The cost would depend on the options chosen. For instance, if it was deemed desirable to retain both spherical and polar drive capability (by adding an equatorial beam), the committee assumes the cost would be in the hundreds of millions of dollars. On the other hand, repositioning the existing beams would presumably cost much less but would narrow the options available to researchers.

The reader should note that the capital cost estimates presented in Tables 4.4 and 4.5 are early-stage estimates and that such estimates for future technology facilities often prove to be underestimates.

THE NEED FOR A NATIONAL INERTIAL FUSION ENERGY R&D PROGRAM

In addition to target science, there are other deep science issues embedded in what is usually labeled "technology" (e.g., chambers) involving a broad range of scientific disciplines, including nuclear and atomic physics, materials and surface science, and engineering science. In the next several years, the IFE program will probably not be involved in engineering development but rather in science and engineering research aimed at determining if feasible solutions exist to the very challenging problems.

An organized program that encompasses all technology options most effectively determines the roadmap to an IFE DEMO plant. Only such a program will have a broad enough view to ultimately identify the most promising IFE DEMO design(s).

The committee recognizes how challenging and complex the unresolved issues are and how much remains to be accomplished and understood if IFE is to become a practical energy source. Each potential driver and target combination has advantages and disadvantages, technologies are evolving rapidly, and scientific challenges remain. If the nation intends to establish IFE as part of its energy R&D portfolio, it is clear that both science and technology components must be addressed in an integrated and coordinated effort.

The roadmap concept put forward by this committee carries forward all IFE approaches to some point at which off-ramp or continuation decisions are made. Should the NIF achieve ignition with indirect drive and the nation decide to pursue IFE, the R&D required to pursue IFE as a practical energy option would begin to diverge from the R&D that NNSA is likely to support for stockpile stewardship applications. In this case, a nationally coordinated R&D program for IFE would be needed to pursue a broad-based roadmap. Inertial fusion energy is an integrated concept whose overall probability of success depends on the success of several individual items. If one component fails a physics test or fails to be cost-effective, the system fails, regardless of whether reactor-scale ignition and gain are reached.

There has been considerable discussion within the committee about the timing for—and the extent of—a technology development element (chambers, target fabrication, etc.) as described in Chapter 3, as part of the early phase(s) of the IFE program. The committee recognizes that absent ignition within the physics element of the program, technology would be of limited value as part of the early phase(s) of the IFE program. There are, however, several reasons for establishing a technology element even in the earliest phases of the IFE program.

A program is needed that attempts to answer whether there is any TA that appears to be practical as well as economically viable. Only certain combinations of targets, drivers, and chambers seem to be possible in this sense. While the emphasis today and in the near future should be on scientific issues related to driver and target performance, working only on these problems could easily lead to solutions that are not compatible with practical commercial driver and chamber options. Such a serial approach could lead to dead ends and would also extend the timescale to the possible practical implementation of IFE.

Technology R&D is not done in a vacuum, and certain answers from the technology research will be beneficial to the overall IFE program in its earlier phases. The design of a FTF and a DEMO cannot be accomplished absent critical technology developments even in the conceptual stages. If the IFE program is to continue advancing, there must be supporting technology developments all along the event pathways. And, perhaps most importantly, if there is to be a meaningful IFE program, it is vital that there be a skilled workforce to investigate the myriad of technology problems over the coming decades. These trained technical experts will not be available unless there is meaningful and challenging R&D for them to carry out early on. That will be possible only if there is a long-term sustained technology element included in the IFE program. Such a program element can be enhanced by identifying synergistic opportunities between the magnetic fusion energy and IFE programs and incorporating them in both programs.

Conclusion 4-13: The appropriate time for the establishment of a national, coordinated, broad-based inertial fusion energy program within DOE is when ignition is achieved.

Conclusion 4-14: There is a compelling need for a sustained, long-term engineering science and technology component in a national inertial fusion energy program.

Such a program would require a sustained effort that is initially devoted primarily to an improved understanding of target physics, particularly the relationship between absorbed energy and gain. Once the target physics is understood, modest gain has been achieved, and there is confidence that reactor-scale gain can be achieved, funding would then be ramped up and devoted primarily to technology development of the three TAs, including target manufacture, driver modules, chamber design, and materials. TA (driver) down-select should occur as part of the technology development phase. The committee's order-of-magnitude estimate for accomplishing this in a two-step approach is given in Table 4.4.

Recommendation 4-9: An engineering science and technology development component should be included in a national inertial fusion energy program.

Conclusion 4-15: The National Ignition Facility (NIF), designed for stockpile stewardship applications, is also of great potential importance for advancing the technical basis for inertial fusion energy (IFE) research.

For a national IFE program, NIF can be utilized for ignition optimization and for demonstration of reactor-scale gain and of reactor-scale gain with more cost-effective targets, as the target physics of direct drive and indirect drive advance technically. Furthermore, modification of NIF to accommodate polar direct drive would not preclude further experiments with indirect drive. This appears to be consistent with the NNSA strategy following completion of the NIC.[15]

Recommendation 4-10: Planning should begin for making effective use of the National Ignition Facility as one of the major program elements in an assessment of the feasibility of inertial fusion energy.

With the approach described here, there needs to be a serious discussion about how such a program should be managed. Certainly it is the prerogative and responsibility of DOE to make such a decision. However, in the interests of cost-effectiveness and efficiency, the committee is of the opinion that a single programmatic office should be established. The committee recognizes that, for an extended period, some overlap will likely continue with programs needed for stockpile stewardship, but that an early effort will be required to facilitate the transition to a national IFE program and to minimize the potential for some overlap.

Conclusion 4-16: At the present time, there is no single administrative home within the Department of Energy that has been invested with the responsibility for administering a national inertial fusion energy R&D program.

Recommendation 4-11: In the event that ignition is achieved on the National Ignition Facility or another facility, and assuming that there is a federal commitment to establish a national inertial fusion energy R&D program, the Department of Energy should develop plans to administer such a national program (including both science and technology research) through a single program office.

---

[15] J. Quintenz, NNSA, Presentation to the committee on February 22, 2012, and LLNL, 2012, "Polar Drive Ignition Campaign Conceptual Design," LLNL TR-553311, submitted to NNSA in April 2012 by LLNL and revised and submitted to NNSA by LLE in September 2012.

It is expected that this would facilitate the management and planning of a focused, coordinated, cost-effective national program, the development of the necessary technologies, and eventual down-selection among driver options and target designs. A single program office would also facilitate the transition of the national IFE program from a science- and technology-based R&D program in the near term to an engineering-based development program in the long term.

In the interim, while IFE is being funded by several offices, it is important to utilize to the maximum extent possible existing facilities in the NNSA and Office of Fusion Energy Sciences programs to minimize costs as much as possible. This will also be true if a national IFE program is established.

Appendixes

A

The Basic Science of Inertial Fusion Energy

The aim of inertial confinement fusion (ICF) is to ignite a target containing compressed fusion fuel—deuterium (heavy hydrogen) and tritium (super-heavy hydrogen)—so that it will burn (react) significantly before the target blows itself apart. Clearly, if this is to be of use for energy production, the energy required to initiate the burn must be significantly less than the energy released by the fusion reactions. Furthermore the energy release of the target must also be sufficiently small that it can be contained and converted into useful power. This appendix outlines the basic physics of the process as it is currently envisaged.

The thermonuclear reaction between deuterium and tritium (DT) yields helium (an alpha particle) and a neutron. The neutron is used to "breed" tritium from lithium in a secondary reaction (see Figure A.1). The energy released is huge: burning only 12 mg of a 50-50 DT mixture yields 4.2 GJ of energy, equivalent to 1 ton of TNT.

In a DT plasma at temperatures over about 50 million degrees, random collisions of D and T produce more energy via the fusion reaction than is radiated away by photons. This is the expected initiation temperature for fusion burn—typically the plasma would then heat itself to above 200 million degrees while burning. The reaction rate per particle depends on temperature and density. At 200 million degrees the reaction rate per particle is $5.2 \times 10^7 \rho s^{-1}$, where r is the DT mixture's mass density in grams per cubic centimeter. The disassembly time of an isothermal sphere is roughly $R/(3C_s)$, where R is the radius and C_s is the speed of sound—at 200 million degrees, C_s is roughly 10^8 cm/s. Thus (very approximately) the areal density, ρR, must be >3-7 g/cm² in order to get a significant proportion of the

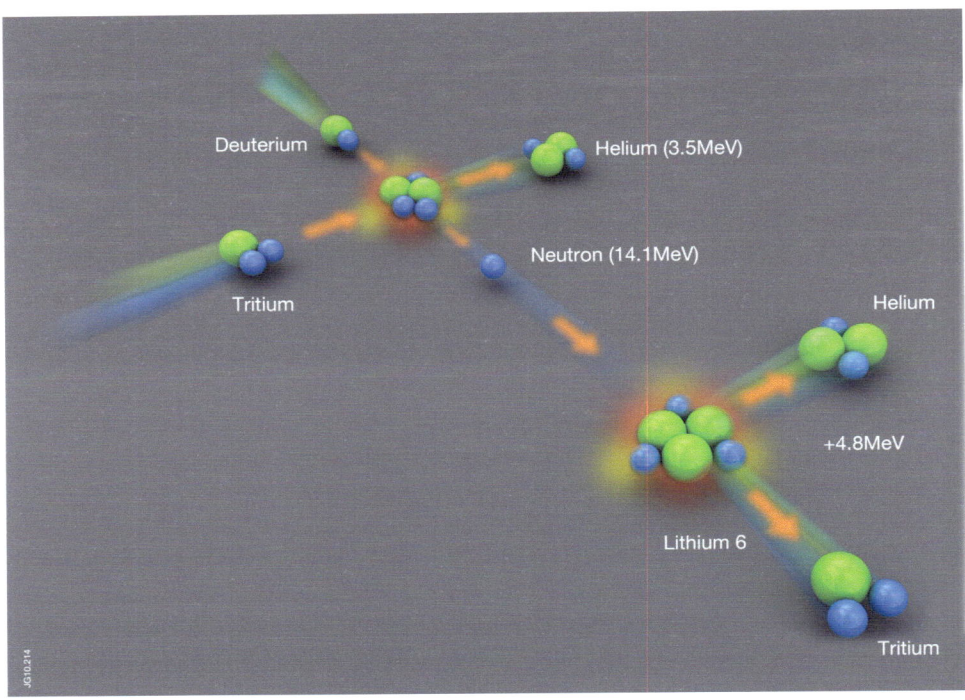

FIGURE A.1 The deuterium-tritium fusion reaction and the tritium breeding reaction from lithium 6. SOURCE: Steve Cowley, United Kingdom Atomic Energy Authority, and Imperial College London.

nuclei to react in the disassembly time. At DT liquid density this would require a sphere 10-30 cm in radius and a huge release of energy. To keep the energy to initiate fusion small and the energy released manageable, a small sphere (weighing a few milligrams) must be used. This requires compression. The areal density rises during compression (at fixed mass, $\rho R \propto R^{-2}$) until it reaches a substantial fraction of fusion-relevant levels (of order 3-7 g/cm^2). For 3 mg of solid/liquid DT an increase in the density of order one thousand is needed.

In most ICF schemes, a shell of cryogenic deuterium and tritium fuel is accelerated inward and compressed by the reaction force from an ablating outer shell. The ablating outer shell is heated either by direct laser irradiation (called "direct drive") or by the X-rays produced by heating a high-Z enclosure (hohlraum) that surrounds the fuel target (called "indirect drive"). The hohlraum in indirect drive schemes may be driven (heated) by lasers, particle beams, or pulsed power systems. During compression the fuel is kept as cold as possible to minimize the work needed for compression. At stagnation, a central hot spot enclosing a few percent of the total mass is heated and ignited. Ignition occurs when the alpha-particle heating of the hot spot exceeds all the energy losses. Ignition triggers a runaway process (the thermonuclear instability), resulting in a large amplification of the hot spot energy. If the inertia of the surrounding dense DT shell confines the ignited hot spot pressure long enough, the thermonuclear burn will propagate from the central hot spot to the dense shell and the entire fuel mass will burn. The burn is driven by the fusion alpha particles depositing their energy in the cold dense fuel. The burn lasts until the target disassembles, and the fuel burn-up fraction increases with the shell areal density.

Compressing a target to ignition conditions is very challenging and has yet to be fully realized in experiments, although major advances have been made. Drivers must deliver very uniform ablation; otherwise the target is compressed asymmetrically. Asymmetric compression excites strong Rayleigh-Taylor instabilities that spoil compression and mix dense cold plasma with the less dense hot spot. Preheating of the target can also spoil compression. For example, mistimed driver pulses can shock heat the target before compression. Also, interaction of the driver with the surrounding plasma can create fast electrons that penetrate and preheat the target.

A widely used parameter to assess the performance of an ICF target is the target gain, G, representing the ratio of the fusion energy output to the driver energy entering the target chamber. Clearly a high gain is desirable for fusion energy and must remain a central focus of any IFE program.

The fraction of driver energy that couples to the fusion fuel contained in the target is typically small—a few percent—but the fusion gain can still be substantial. In a National Ignition Facility (NIF) indirect-drive ignition target driven by ~1 MJ of UV laser light into the hohlraum, the shell of fuel implodes with an expected kinetic energy of about 15-20 kJ. Roughly half of that energy (7-10 kJ) is used to heat up the hot spot and the other half to compress the surrounding shell. If the fusion yield (alpha and neutron energy) is 1 MJ (i.e., G = 1), the hot spot energy is amplified 100× by the thermonuclear instability. At 1 MJ fusion yield, the alpha particles have deposited 200 kJ of energy into the hot spot and surrounding fuel, about 20 times the energy provided by the compression of the hot spot. The thermonuclear burn stays localized near the hot spot and propagates within about 5 times the initial hot spot mass (partial burn). If the burn propagates through the entire DT mass, the gain of a NIF target will exceed ~10 (full burn and 10 MJ yield). While a NIF implosion yielding G >> 1 would elucidate many aspects of the ignition and basic burn physics, a gain of G ≥ 10 is required for demonstrating full burn propagation over the inertial confinement time of the compressed shell (i.e., fuel burnup fraction compatible with the fuel inertia).

While the target gain can be used to validate the target physics, a new parameter is required for assessing the viability of a fusion energy system. The so-called "Engineering Q," or "Q_E," is often used as a figure of merit for a power plant. It represents the ratio of the total electrical power produced to the (recirculating) power required to run the plant—i.e., the input to the driver and other auxiliary systems. Clearly $Q_E = 1/f$, where f is the recycling power fraction (see Figure A.2). Typically $Q_E \geq 10$ is required for a viable electrical power plant. For a power plant with a driver wall-plug efficiency η_D, target gain G, thermal-to-electrical conversion efficiency η_{th}, and blanket amplification A_B (the total energy released per 14.1 MeV neutron entering the blanket via nuclear reactions with the structural, coolant, and breeding material), the engineering Q is $Q_E = \eta_{th}\eta_D A_B G$ (see Figure A.2). An achievable value of the blanket amplifications and thermal efficiency might be $A_B \sim 1.1$ and $\eta_{th} \sim 0.4$ and should be largely independent of the driver. Therefore, the minimum required target gain is inversely proportional to the driver efficiency. For a power plant with a recirculating power f = 10 percent ($Q_E = 10$), the required target gain is G = 150 for a 15 percent efficient driver and G = 320 for a 7 percent efficient driver.

Energy gain does not, of course, guarantee commercial viability. Key challenges remain even after high gain is achieved. These are discussed in detail elsewhere in this report, but they include:

- *Low-cost targets.* For example, a target producing a fusion energy, E_D, of 200 MJ could make net electricity, $E_{grid} \sim 80$ MJ ~ 22 kWh, or about $1 worth of electricity at current prices. The target cost should be some small fraction of this.
- *Repetitive ignition of targets.* To produce a gigawatt of electrical power, targets with $E_D = 200$ MJ must be ignited roughly 12 times a second.
- *Reliable target chamber and blanket to extract power and breed tritium.* This is a challenge shared with magnetic fusion.

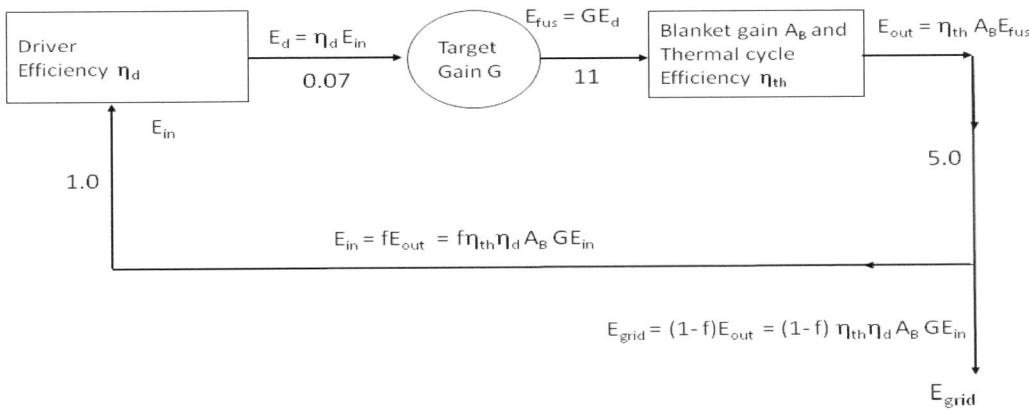

FIGURE A.2 Schematic energy flow in an inertial fusion power plant. Note the "Engineering Q" is defined as $Q_E = 1/f$. The numbers beside the arrows indicate the proportionality of the energy flows. Tritium breeding (discussed in Chapter 3) is excluded from this diagram for simplicity.

B

Statements of Task

The statements of task for both the committee's final report and its interim report are shown below. The scope of the final report was intended to be much broader than that of the interim report. The statement of task for the separate and supporting study by the Panel on the Assessment of Inertial Confinement Fusion (ICF) Targets is also shown.

FOR THE COMMITTEE ON THE PROSPECTS FOR INERTIAL CONFINEMENT FUSION ENERGY SYSTEMS

The statement of task for the committee is as follows:

The Committee will prepare a report that will:

- Assess the prospects for generating power using inertial confinement fusion;
- Identify scientific and engineering challenges, cost targets, and R&D objectives associated with developing an IFE demonstration plant; and
- Advise the U.S. Department of Energy on its development of an R&D roadmap aimed at creating a conceptual design for an inertial fusion energy demonstration plant.

The Committee will also prepare an interim report to inform future year planning by the federal government.

A Panel on Fusion Target Physics with access to classified information as well as controlled-restricted unclassified information will serve as a technical resource to the committee and will describe, in a report containing only publicly accessible information, the R&D challenges to providing suitable targets on the basis of parameters established and provided by the Committee. The Panel will also assess the current performance of various fusion target technologies.

FOR THE PANEL ON THE ASSESSMENT OF
INERTIAL CONFINEMENT FUSION (ICF) TARGETS

The statement of task for the supporting panel is as follows:

A Panel on Fusion Target Physics ("the Panel") will serve as a technical resource to the Committee on Inertial Confinement Energy Systems ("the Committee") and will prepare a report that describes the R&D challenges to providing suitable targets, on the basis of parameters established and provided to the Panel by the Committee.

The Panel on Fusion Target Physics will prepare a report that will assess the current performance of fusion targets associated with various ICF concepts in order to understand:

1. The spectrum output;
2. The illumination geometry;
3. The high-gain geometry; and
4. The robustness of the target design.

The Panel will also address the potential impacts of the use and development of current concepts for Inertial Fusion Energy on the proliferation of nuclear weapons information and technology, as appropriate. The Panel will examine technology options, but will not provide recommendations specific to any currently operating or proposed ICF facility.

C

Agendas for Committee Meetings and Site Visits

FIRST MEETING
KECK CENTER, WASHINGTON, D.C.
DECEMBER 16-17, 2010

Thursday, December 16, 2010

Closed Session

8:30 am	Committee discussion	*Ron Davidson and Jerry Kulcinski, Co-Chairs*
Noon	Working lunch (continued discussion)	*Committee*

Open Session

1:00 pm	Welcome	*Ron Davidson and Jerry Kulcinski, Co-Chairs*
1:15	Perspectives from the DOE Office of Science	*Steve Koonin, DOE*
1:45	Discussion	
2:00	Perspectives from NNSA Stockpile Stewardship	*Chris Deeney, NNSA*
2:20	Discussion	
2:30	Perspectives from the DOE Office of Fusion Energy Science	*Ed Synakowski and Mark Koepke, OFES*

3:00	Discussion	
3:15	Break	
3:30	Findings from the 2003 FESAC report: *A Plan for the Development of Fusion Energy*	Robert Goldston, PPPL, and Michael Campbell, Logos Technologies
4:00	Discussion	
4:15	Findings from the 2004 FESAC report: *Review of the Inertial Fusion Energy Program*	Rulon Linford
4:45	Discussion	
5:00	Public comment session	Audience
6:00	Adjourn meeting for the day	

Closed Session

6:30	Working dinner	

Friday, December 17, 2010

Closed Session

7:30 am	Breakfast	
8:30	Committee discussion	Co-Chairs

Open Session

9:00	Perspectives from the DOE Office of Science	Bill Brinkman, DOE
9:30	Discussion	
9:45	Perspectives from NNSA Defense Programs	Donald Cook, NNSA
10:15	Discussion	
10:30	Break	
10:45	Challenges to Developing an ICF-based Energy Source	Harold Forsen
11:15	Discussion	
11:30	Perspectives from the Office of Science and Technology Policy	Steve Fetter, OSTP
11:45	General discussion	

APPENDIX C

Closed Session

12:15 pm	Working lunch	
1:00	Committee discussion	*Committee*
3:00	Adjourn	

SECOND MEETING
SAN RAMON, CALIFORNIA
JANUARY 29-31, 2011

Saturday, January 29, 2011

Open Session

8:00 am	Welcome and Opening Remarks	*Ron Davidson and Jerry Kulcinski, Co-Chairs*
8:15	Laser-Driven Inertial Fusion Energy, Indirect-Drive Targets (including Q&A) Lawrence Livermore National Laboratory	*Michael Dunne, Edward Moses, Jeff Latkowski, and Tom Anklam, LLNL*
10:15	Break	
10:30	Laser-Driven Inertial Fusion Energy, Direct-Drive Targets (including Q&A) University of Rochester	*Robert McCrory, Stanley Skupsky, and Jonathan Zuegel, LLE*

Closed Session

12:30 pm	Working lunch: preparation of questions for speakers from morning sessions

Open Session

1:00	Krypton-Fluoride-Driven Inertial Fusion Energy (including Q&A) Naval Research Laboratory	*John Sethian and Stephen Obenschain, NRL*
3:00	Break	
3:15	Ion-Beam-Driven Inertial Fusion Energy (including Q&A) Lawrence Berkeley National Laboratory	*Grant Logan, LBNL*

Closed Session

4:45   Discussion and preparation of questions
         for speakers from afternoon sessions

Open Session

5:00   Question and answer session with speakers
         on all driver concepts
6:00   Adjourn open session

Closed Session

6:00   Committee discussion
9:00 pm   Adjourn for the day

Sunday, January 30, 2011

Closed Session

7:30 am   Breakfast

Open Session

8:00   Pulsed-Power Inertial Fusion Energy and   *Michael Cuneo and*
         Targets   *Mark Herrmann, SNL*
         (including Q&A)
         Sandia National Laboratories

Closed Session

9:30   Discussion and preparation of questions
         for morning speaker

Open Session

9:45 am	Questions and answer session with morning speaker	
10:00	Perspectives from Los Alamos National Laboratory (including Q&A)	*Juan Fernández, LANL*
10:45	Overview of IFE Target Designs (including Q&A)	*John Perkins, LLNL*
11:45	Break for lunch	
Noon	Overview of Chamber and Power Plant Designs for IFE (including Q&A)	*Wayne Meier, LLNL*
1:00 pm	Target Fabrication and Injection (including Q&A)	*Dan Goodin, General Atomics*
2:00	Perspective of Stephen Bodner (including Q&A)	*Stephen Bodner*
2:45	General question and answer period	
3:15	Public comment session	*All*
4:15	Adjourn open session	

Closed Session

4:15	Committee discussion
8:30 pm	Adjourn for the day

Monday, January 31, 2011

Open Session

8:00 am	Site Visit: Lawrence Livermore National Laboratory
12:30 pm	Lunch at LBNL
1:30	Site Visit: Lawrence Berkeley National Laboratory
4:00	Adjourn meeting

THIRD MEETING
ALBUQUERQUE, NEW MEXICO
MARCH 29-31 AND APRIL 1, 2011

Tuesday, March 29, 2011

Closed Session

7:00 pm	Inertial Confinement Fusion and Inertial Fusion Energy Tutorial (committee only)	*Steve Cowley and Riccardo Betti, Committee members*
9:00	Adjourn for the day	

Wednesday, March 30, 2011

Closed Session

8:00 am	Welcome and opening remarks: Plans and goals for the meeting	*Ron Davidson and Jerry Kulcinski, Co-Chairs*
8:30	Balance and composition discussion for new members	*David Lang, NRC*
8:45	Break	

Open Session

9:00	Welcome and opening remarks	*Ron Davidson and Jerry Kulcinski, Co-Chairs*
9:05	The National Ignition Campaign	*John Lindl, LLNL*
10:00	Discussion	
10:15	Role of the National Ignition Facility Beyond the National Ignition Campaign: NNSA Perspective	*Chris Deeney, NNSA*
10:45	Discussion	
11:00	LIFE Delivery Plan	*Mike Dunne and others, LLNL*
Noon	Discussion	

Closed Session

12:15 pm	Lunch	*Committee only*

Open Session

1:00	Fast Ignition for Inertial Fusion Energy	*Richard Freeman,* *Ohio State University*
1:45	Discussion	
2:00	Adjourn open session for the day	

Closed Session

2:15	Discussion with ICF Target Physics Panel Chair	*John Ahearne, Chair,* *Target Physics Panel* *(by teleconference)*
3:15	Committee discussion	
8:30 pm	Adjourn for the day	

Thursday, March 31, 2011

Closed Session

7:30 am   Breakfast

Open Session

8:00	Magnetized Target Fusion	*Glen Wurden, LANL, and* *Irv Lindemuth, University* *of Nevada at Reno*
8:45	Discussion	
9:00	Chamber Materials Challenges for Inertial Fusion Energy	*Steve Zinkle, ORNL*
10:00	Discussion	
10:15	Break	
10:30	Lessons in Engineering Innovation	*Elon Musk,* *SpaceX, Tesla Motors,* *Solar City* *(by videoconference)*
11:00	Public comment session	
Noon	Adjourn open session	

Closed Session

Noon	Lunch	*Committee only*
1:00 pm	Committee discussion	
8:30	Adjourn for the day	

Friday, April 1, 2011
Site Visit to Sandia National Laboratories

Closed Session

8:00 am	Remarks on Sandia and IFE	*Steve Rottler, SNL*
8:30	Various presentations	
10:00	Break	
10:25	Tour of the Z facility	
11:00	Mykonos facility	
Noon	Adjourn meeting	

FOURTH MEETING
ROCHESTER, NEW YORK
JUNE 15-17, 2011

Wednesday, June 15, 2011

Closed Session

8:30 am	Welcome and opening remarks	*Ron Davidson and Jerry Kulcinski, Co-Chairs*
8:45	Break	

Open Session

9:00	Welcome and opening remarks	*Ron Davidson and Jerry Kulcinski, Co-Chairs*
9:05	Inertial Fusion Energy: Activities and Plans in the UK and EU	*John Collier, UK Science and Technology Facilities Council*
10:15	Discussion	
10:35	Break	

10:50	Inertial Fusion Energy: Activities and Plans in Japan	*Hiroshi Azechi, Institute of Laser Engineering, Osaka University*
Noon	Discussion	
12:20 pm	Lunch	
1:00	Integrated Design of a Laser Fusion Target Chamber System	*John Sethian, Naval Research Laboratory*
2:00	Discussion	
2:20	Adjourn open session for the day	

Closed Session

2:30	Discussion
8:30 pm	Adjourn for the day

Thursday, June 16, 2011

Closed Session

8:00 am	Breakfast

Open Session

8:30	Nuclear Power Plant Financing	*Philip M. Huyck, Encite, LLC (formerly of Credit Suisse First Boston & Trust Company of the West)*
9:30	Discussion	
9:45	Inertial Fusion Energy: Activities and Plans in China	*Zhang Jie, Shanghai Jiao Tong University*
11:00	Discussion	
11:20	Public comment session	
11:30	General discussion with all speakers	*Committee and speakers*
Noon	Adjourn open session	

Closed Session

Noon	Lunch	*Committee only*
1:00 pm	Discussion with ICF Target Physics Panel Chair	*John Ahearne, Chair, Target Physics Panel*
2:00	Continued discussion	
8:30 pm	Adjourn for the day	

Friday, June 17, 2011
Site Visit to the Laboratory for Laser Energetics

Closed Session

8:00 am	Discussion	*All*
9:30	Break and gather for site visit	

Open Session

9:45	LLE Overview	*R.L. McCrory, LLE*
10:15	Site tours and posters	

- Break panel into three groups, each with a primary tour guide. Tour guides: R.L. McCrory, D.D. Meyerhofer, and P. McKenty
- Three stations, each with two posters and a facility presenter (~1/2 hour at each station)
    — OMEGA
    - S. Morse
    - Poster on cryogenic target performance and polar drive—V. Goncharov
    - Poster on OMEGA as a user facility—J. Soures
    — OMEGA EP
    - D. Canning
    - Poster on fast/shock ignition—W. Theobald
    - Poster on new technologies for EP—J. Zuegel
    — OMAN
    - A. Rigatti
    - Poster on high damage threshold coatings—J. Oliver
    - Poster on diffractive optics—T. Kessler

Noon   Adjourn site visit, adjourn meeting

APPENDIX C

FIFTH MEETING
WASHINGTON, D.C.
OCTOBER 31-NOVEMBER 2, 2011

Monday, October 31, 2011

Closed Session

8:30 am   Committee discussion

Open Session

10:15   Welcome and opening remarks   *Ron Davidson and*
                                                Jerry Kulcinski, Co-Chairs
10:20   Heavy Ion Inertial Fusion Energy: *Boris Sharkov,*
            Activities and Plans in Europe and Russia   *FAIR GmbH*
11:20   Discussion
11:40   Public comment session

Closed Session

Noon   Lunch   *Committee only*

Open Session

1:00 pm Mass Manufacturing of Targets   *Abbas Nikroo,*
                                                General Atomics
2:00   Discussion
2:30   A Perspective on Licensing of Inertial *Dick Meserve,*
            Fusion Power Plants   *Carnegie Institute for*
                                                Science
3:00   Discussion

Closed Session

9:00 pm Adjourn for the day

Tuesday, November 1, 2011

Closed Session

Open Session

10:45 am	A Perspective on Safety Issues of an Inertial Fusion Power Plant	*Kathy McCarthy, Idaho National Laboratory*
11:15	Discussion	
11:30	Public comment session	

Closed Session

6:30 pm   Adjourn for the day

Wednesday, November 2, 2011
Site Visit to Laser Fusion Facilities, Naval Research Laboratory

Open Session

9:00 am	Gathering and introductions	
	Overview of the NRL Laser Fusion Program	*Stephen Obenschain, Victor Serlin, and John Sethian*
9:45	Tours of Nike and Electra KrF Laser Facilities	
	Tour of Nike Target Facility	*Yefim Aglitskiy, Max Karasik, and Jim Weaver*
	Tour of Nike Laser Facility	*David Kehne and Steve Terrell*
	Tour of Electra Facility	*Frank Hegeler, Matt Myers, and Matt Wolford*
11:15	Discussion (with light lunch)	
11:45	Adjourn	

Appendix C

SIXTH MEETING
SAN DIEGO, CALIFORNIA
FEBRUARY 22-23, 2012

Wednesday, February 22, 2012

Closed Session

8:00 am	Welcome	*Ron Davidson and Jerry Kulcinski, Co-Chairs*
8:05	Discussion of Business: Status of the Study	*David Lang, Staff*
8:30	Report from the ICF Target Physics Panel	*John Ahearne, Chair*

Open Session

9:00	Status of the National Ignition Campaign and Plans Post-FY2012	*Jeffrey Quintenz, NNSA*
9:25	Discussion	
9:40	Status of the National Ignition Facility, Plans for the Facility Post-FY2012, and the LIFE Project	*Mike Dunne, LLNL*
10:05	Discussion	
10:20	Public comment session	
10:40	Adjourn open session	

Closed Session

10:45	Discussion of final report
Noon	Working lunch

Open Session

12:30 pm	Tour of General Atomics target fabrication facilities	*All*
2:00	Adjourn tour and open session	

Closed Session

2:05	Continued discussion of final report
6:00 pm	Adjourn for the day

Thursday, February 23, 2012

Closed Session

8:00 am	Continued discussion of final report	
4:00 pm	Discussion of plan to complete report	*All*
5:00 pm	Adjourn meeting	

D

Agendas for Meetings of the Panel on the Assessment of Inertial Confinement Fusion (ICF) Targets

FIRST MEETING
WASHINGTON, D.C.
FEBRUARY 16-17, 2011

Call to order and welcome
John Ahearne, Chair

Overview of the study task and origins and the National Academies' study process
Sarah Case, Study Director; John Ahearne, Chair

IFE committee briefing to the panel on expectations
Gerald Kulcinski, Inertial Fusion Energy Committee Co-Chair

Review of charge to the panel, the U.S. Department of Energy's interests in the committee and panel reports, and nuclear weapons proliferation risks for an inertial fusion energy program
David Crandall, Office of the Under Secretary for Science, U.S. Department of Energy

Indirect drive target physics at the National Ignition Facility (NIF)
John Lindl, Lawrence Livermore National Laboratory (LLNL)

Direct drive target physics at the Naval Research Laboratory (NRL)
Andrew Schmitt, NRL

Direct drive target physics at NIF
David Meyerhofer, Laboratory for Laser Energetics

Heavy ion target physics
John Perkins, LLNL

Z-pinch target physics
Mark Herrmann, Sandia National Laboratories (SNL)

Non-proliferation considerations associated with inertial fusion energy
Raymond Jeanloz, University of California, Berkeley

SECOND MEETING
PLEASANTON, CALIFORNIA
APRIL 6-7, 2011

Welcome and call to order
John Ahearne, Chair

System considerations for IFE
Tom Anklam, LLNL

Overview of laser inertial fusion energy system and key considerations for IFE targets
Michael Dunne, LLNL

THIRD MEETING
ALBUQUERQUE, NEW MEXICO
MAY 10-11, 2011

Welcome and call to order
John Ahearne, Chair

Inertial confinement fusion (ICF) targets at Los Alamos National Laboratory
Juan Fernandez, LANL

Design and simulation of magnetized liner inertial fusion targets
Steve Slutz, SNL

Appendix D

**FOURTH MEETING
ROCHESTER, NEW YORK
JULY 6-8, 2012**

Welcome and call to order
John Ahearne, Chair

Welcome and overview of Laboratory for Laser Energetics (LLE) ICF program
Robert McCrory, LLE

Direct-drive progress on OMEGA
Craig Sangster, LLE

Polar drive target design
Radha Bahukutumbi, LLE

Facilitating NIF for polar drive
David Meyerhofer, LLE

Fast and shock ignition research
David Meyerhofer, LLE

LPI issues for direct drive
Dustin Froula and Jason Myatt, LLE

Heavy ion target design
B. Grant Logan, LBNL

Discussion of LIFE targets and program
Michael Dunne, LLNL

Technical feasibility of target manufacturing
Abbas Nikroo, General Atomics

**FIFTH MEETING
WASHINGTON, D.C.
SEPTEMBER 20-22, 2012**

Welcome and call to order
John Ahearne, Chair

Development of the technologies for laser fusion direct drive
John Sethian, NRL

Overview of current NRL program for ICF/IFE
Steve Obenschain and Andrew Schmitt, NRL, and Frank Hegeler, Commonwealth Technology at NRL

Overview of LPI physics and LANL understanding
David Montgomery, LANL

Understanding of LPI and its impact on indirect drive
Mordechai Rosen, LLNL

Assessment of understanding of LPI for direct drive (solid-state)
Dustin Froula, LLE

Assessment of understanding of LPI for direct drive (KrF)
Andrew Schmitt, NRL

State of the art for LPI simulation
Denise Hinckel, LLNL

E

Bibliography of Previous Inertial Confinement Fusion Studies Consulted by the Committee

"2002 Fusion Summer Study Report." 2003. Snowmass, Colorado. July 8-19.

R. Davidson, B. Ripin, M. Abdou, et al. 1994. Fusion Energy Advisory Committee (FEAC): Panel 7 report on inertial fusion energy, *Journal of Fusion Energy* 13(2/3): 233-260.

Executive Office of the President, President's Council of Advisors on Science and Technology. 2010. "Report to the President on Accelerating the Pace of Change in Energy Technologies Through an Integrated Federal Energy Policy." November.

Fusion Energy Sciences Advisory Committee. 1999. "Opportunities in the Fusion Energy Sciences Program." June.

———. 1999. "Report of the FESAC Panel on Priorities and Balance." September.

———. 2001. "Review of the Fusion Theory and Computing Program." August.

———. 2002. "Report of the Fusion Energy Sciences Advisory Committee Burning Plasma Strategy Panel: A Burning Plasma Program Strategy to Advance Fusion Energy." September.

———. 2003. "Report of the Fusion Energy Sciences Advisory Committee Fusion Development Path Panel: A Plan for the Development of Fusion Energy." March.

———. 2004. "Review of the Inertial Fusion Energy Program." March.

———. 2005. "Scientific Challenges, Opportunities and Priorities for the U.S. Fusion Energy Sciences Program." April.

———. 2009. "Panel on High Energy Density Laboratory Plasmas: Advancing the Science of High Energy Density Laboratory Plasmas." January.

National Research Council. 1986. *Review of the Department of Energy's Inertial Confinement Fusion Program.* Washington, D.C.: National Academy Press.

———. 1990. *Review of the Department of Energy's Inertial Confinement Fusion Program.* Washington, D.C.: National Academy Press.

———. 1997. *Review of the Department of Energy's Inertial Confinement Fusion Program: The National Ignition Facility.* Washington, D.C.: National Academy Press.

———. 2003. *Frontiers in High Energy Density Physics: The X-Games of Contemporary Science.* Washington, D.C.: The National Academies Press.

———. 2004. *Burning Plasma: Bringing a Star to Earth.* Washington, D.C.: The National Academies Press.

———. 2007. *Plasma Science: Advancing Knowledge in the National Interest.* Washington, D.C.: The National Academies Press.

J. Sheffield, M. Abdou, R. Briggs, et al. 1999. Report of the FEAC inertial fusion energy review panel: July 1996, *Journal of Fusion Energy* 18(4): 195-211.

NOTE: For brevity, the committee presents here only studies it consulted that were produced by the National Research Council and federal advisory committees. A full list of materials consulted by the committee is available through the National Academies' Public Access Records Office.

F

Foreign Inertial Fusion Energy Programs

Countries other than the United States and consortia of countries are seeking to attain fusion energy. These facilities and programs are briefly described in this appendix.

EUROPEAN UNION—HIGH POWER LASER ENERGY RESEARCH (HiPER)

The High Power Laser Energy Research (HiPER) project is an international collaborative research activity to design a high-power laser fusion facility capable of "significant energy production."[1] It is funded by 10 funding agency partners in the European Union (from the United Kingdom, France, the Czech Republic, Greece, Spain, and Italy) and has 17 institutional partners. A coordinated science and technology effort to achieve HiPER exists between major laser laboratories such as Laser MégaJoule (LMJ), the PETawatt Aquitaine Laser (PETAL), Orion, the Extreme Light Infrastructure (ELI), and the Prague Asterix Laser System (PALS), with each lab investigating discrete elements of interest.

The driver for HiPER consists of diode pumped solid-state lasers (DPSSLs). Their preliminary design has not yet specified a particular DPSSL material, but a few are under consideration at this time, such as cryocooled Yb:CaF_2, Yb:YAG, and ceramic Yb:YAG. These materials can be made in large sizes, easily scaled, and have a wide industrial base on which to draw on from the EU countries.

---

[1] Available at http://www.hiper-laser.org/overview/hiper.asp.

Although other methods are under consideration, HiPER appears to favor the direct drive, shock ignition method. The project is collaborating with universities on the development of technologies for fast ignition. HiPER appears to have no intention of pursuing indirect drive ignition, possibly, at least in part, because French law forbids use of military program data for civilian use. The U.K.'s Atomic Weapons Establishment has been working with the United States on indirect drive at the National Ignition Facility (NIF).

The preliminary design for the ignition target for HiPER uses an aluminum shell containing deuterium-tritium (DT) ice and vapor; a gain greater than 100 is desired for commercial inertial fusion energy (IFE) purposes. Mass production, cryolayering, and chamber injection of these targets are currently under study by Micronanics, General Atomics, and laboratories in the Czech Republic. Much of the design of European approaches to IFE is being done using DUED, a code developed in Italy, and MULTI, a code developed in Spain.

A two-stage development approach to the HiPER chamber is under consideration. The first stage would be a technology integration demonstration, while the next stage would be an IFE reactor. A "consumable" first wall concept is being studied wherein the damaging effects of debris and reaction products on the first wall are mitigated. One consumable wall concept involves gas-filled removable tiles as a modular solution to this problem. Partnerships with the magnetic fusion energy (MFE) community could be of interest for solving these challenges, which are not unique to IFE.

A 3- to 5-kJ laser unit representative of a larger modular scheme for HiPER is currently under development by four European Union teams. The goal of this research thrust is to have a 10 percent efficient laser capable of reaching 1 MJ of energy at 10 Hz.

The timeline for the entire HiPER project begins with a technological development and risk reduction phase from the present to approximately 2020; a design, build, and test phase from approximately 2017 to 2029; and, finally, a reactor design phase from approximately 2025 to 2036. These activities are all intended to be done at a single site to reduce costs and redundancies. During this time, it is anticipated that NIF will have achieved ignition, and that HiPER will have received some business investment.

See the Chapter 2 section "The Global R&D Effort on Solid-State Lasers for IFE Drivers" for more information on laser development in Europe.

FRANCE—LASER MEGAJOULE (LMJ)

The Centre lasers intenses et applications (CELIA), centered at the University of Bordeaux, organizes and administers a collaboration among French academic institutions, the Commissariat à l'énergie atomique et aux énergies alternatives

(CEA), and several other European laser collaborations. It attempts to develop relevant industrial connections for all purposes in the Bordeaux area. CELIA is heavily involved in the HiPER project. It is also a very active collaborator with other nations such as Japan and the United States on laser IFE research and with other large programs such as ITER for fusion-related materials research.

The French IFE effort other than HiPER is the Laser MégaJoule (LMJ). LMJ is similar to HiPER in one way and to NIF in a different way. Like NIF, LMJ will use a flashlamp-pumped laser as its driver. LMJ is also structurally very similar to NIF, but with differences in the number of beams and optics. It will use indirect-drive ignition and will produce approximately the same final laser wavelength of 351 nm at a similar maximum energy of 1.8 MJ. LMJ will use indirect drive for the purpose of weapons physics studies, just as NIF does. Though it is associated with the French nuclear weapons program, LMJ is to be used for open research, including IFE, 25 percent of the time, according to the present CEA Commissioner.

Currently, the CEA target laboratory is responsible for all CEA laser target needs. It has no plans to expand its capabilities for mass-production of IFE targets for the time being and will rely on General Atomics for targets for the foreseeable future. The future challenges that LMJ will face in IFE are similar to those facing other programs reliant on indirect drive: building, positioning, and orienting high-velocity targets; managing the large mass present in an indirect-drive-type target; and meeting the higher energy requirement for indirect drive ignition predicted by computer simulation.

It is planned that "first light" experiments from 162 of the intended 240 beams will occur at LMJ in 2014, with ignition experiments starting in 2017. The EU-sponsored petawatt laser arm (PETAL) will also be brought online in parallel with the main LMJ facility.

CHINA—SG-IV

The Chinese IFE program plans to achieve ignition and burn around the year 2020. On the path to that goal, China is updating existing laser research facilities such as SG-II to higher energies and with additional features such as backlighting. The SG-III lamp-pumped Nd:glass facility is also in the process of an upgrade from 8 to 48 beams. The upgrade and construction work will culminate in completion of the 1.5 MJ (351 nm) SG-IV ignition facility.

The laser driver for the SG-IV facility is planned to be Yb:YAG water-cooled DPSSLs operating between 1 and 10 Hz and fired into a 6-m-diameter target chamber. The choice of ignition method and target has not been finalized, though fast ignition is favored with a cone-in-shell target. The indirect drive is still being considered, however. The upgrades to China's existing laser facilities as well as new capabilities are planned to drive target physics and ignition research.

In addition to many experiments devoted to improving understanding of the physics, the Chinese program is developing its own simulation codes. This code suite will be used to design the ignition targets for China's ignition program, and experiments to check simulation designs will be carried out on the upgraded SG-II (SG-IIU) and SG-III lasers.

JAPAN—FIREX AND I-LIFT

The Japanese Fast-Ignition Realization Experiment (FIREX) IFE facility is planning to achieve ignition using the fast ignition technique around 2019. Japan's IFE program is also working on engineering plans for a Laboratory Inertial Fusion Test (i-LIFT) experimental IFE reactor, and it eventually plans to construct an IFE demonstration plant. i-LIFT will feature 100-kJ lasers firing at 1 Hz and a 100-kJ heating laser operating at the same rate. The facility is designed to generate net electricity.

Currently, experimental progress has been focused on fast ignition by performing integrated experiments with the FIREX-I system and the LFEX CPA heating beam. DPSSLs have been selected as the laser driver—Japan believes that its strong semiconductor industry will underpin this choice of technology. It also cites a strong domestic working relationship with the materials and MFE communities. Japan says that most critical elements of IFE reactor construction have been addressed and/or demonstrated, such as mass production of targets and high-speed target injection, magnetic field laser port protection, and liquid first-wall stability.

The current plans for i-LIFT include operation from 2021 to 2032. The Japanese anticipate that their demonstration plant will begin engineering design in 2026, and a single-chamber system will begin to operate in 2029 and will be expanded to a four-chamber commercial plant operating at 1.2 MJ at 16 Hz in 2040.

See Chapter 2 of this report for more information on laser development in Japan.

RUSSIA AND GERMANY—HEAVY ION-BASED INERTIAL FUSION ENERGY

The IFE collaboration between Russia and Germany has chosen heavy ion beams as their driver method, featuring two options. A 10-km radiofrequency linac would be needed for the heavy-ion driver. They are considering both direct fast ignition and indirect drive methods. Bi and/or Pt ion beams would drive either a rotating cylindrical target or a target similar to the capsule-in-hohlraum designs for laser-driven ignition, with a calculated gain of as much as 100. They are also examining the possibility of a fusion-fission-fusion target design using a layer of ^{238}U.

Their proposed target chamber incorporates a two-walled design, with a wetted silicon carbide first wall and a LiPb blanket. The vapor layer generated from the "prepulse" may mitigate a number of potential challenges such as target debris and X-ray damage to the first wall. However, the vapor generated is also a cause for concern in the overall reactor design. The radiation-hydrodynamics code RAMPHY has been used to study the effects of liquid film ablation and radiation transport, as well as other effects of importance to IFE, such as DT capsule implosion and burn, X-ray and charged particle stopping, and neutron deposition.

Experimental work with the synchrotron SIS and the Facility for Antiproton and Ion Research (FAIR) facilities in Germany is intended to investigate beam development and behavior. Other accelerator challenges to overcome include beam wobbling, vacuum instability, and high current injection. The Institute for Theoretical and Experimental Physics Terawatt Accumulator (ITEP-TWAC) project, which will be a main test bed for these issues, is now under construction.

Russia recently announced a project to build a 2.8-MJ laser for inertial confinement fusion and weapons research. The Research Institute of Experimental Physics (RFNC-VNIIEF) will develop the concept.

G

Glossary and Acronyms

GLOSSARY

Ablator: Outermost layer of the target capsule that is rapidly heated and vaporized, compressing the rest of the target.

Adiabat (plasma physics): Determined, for instance, by the ratio of the plasma pressure to the Fermi pressure (the pressure of a degenerate electron gas); used as a measure of plasma entropy.

Blanket: Section of the reactor chamber that serves as the heat transfer medium for the fusion reactor chamber. Some blanket concepts incorporate materials for tritium breeding as well as cooling.

Cryogenic: Involving very low temperatures.

Diode-pumped lasers: Lasers wherein laser diodes illuminate a solid gain medium (such as a crystal or glass).

Direct drive: Inertial confinement fusion (ICF) technique whereby the driver energy strikes the fuel capsule directly.

Driver: Mechanism by which energy is delivered to the fuel capsule. Typical techniques use lasers, heavy-ion beams, and Z-pinches.

Appendix G

Dry wall: Fusion reactor chamber's first wall that employs no liquid or gaseous protection.

Fast ignition: ICF technique whereby the driver gradually compresses the fuel capsule, followed by a high-intensity, ultrashort-pulse laser that strikes the fuel to trigger ignition.

First wall: First surface of the fusion reactor chamber encountered by radiation and/or debris emitted from the target implosion. These walls may vary in composition and execution such as dry, wetted, or liquid jet.

Gain: Ratio of the fusion energy released by the target to the driver energy applied to the target in a single explosion.

Heavy-ion fusion: ICF technique whereby ions of heavy elements are accelerated and directed onto a target.

High average power: Attribute of a driver that, if repeatable, would make it suitable for an IFE-based power plant.

High-energy-density science: Study of the creation, behavior, and interaction of matter with extremely high energy densities.

High repetition rate: Maintaining a high rate for engaging the driver or igniting the target, making it suitable for an IFE-based power plant (e.g., 10 Hz).

Hohlraum: Hollow container in which an ICF target may be placed, whose walls are used to reradiate incident energy to drive the fuel capsule's implosion.

Hydrodynamic instability: Concept in which fluids of differing physical qualities interact, causing perturbations such as turbulence. Examples include Rayleigh-Taylor and Richtmyer-Meshkov instabilities.

Ignition (broad definition): Condition in a plasma when self-heating from nuclear fusion reactions is at a rate sufficient to maintain the plasma's temperature and fusion reactions without having to apply any external energy.

Ignition (IFE): State when fusion gain exceeds unity—that is, when the fusion energy released in a single explosion exceeds the energy applied to the target.

Indirect drive: ICF technique whereby the driver energy strikes the fuel capsule indirectly—for example, by the X-rays produced by heating the high-Z enclosure (hohlraum) that surrounds the fuel capsule.

Inertial confinement fusion (ICF): Concept in which a driver delivers energy to the outer surface of a fuel capsule (typically containing a mixture of deuterium and tritium), heating and compressing it. The heating and compression then initiate a fusion chain reaction.

Inertial fusion energy: Concept whereby ICF is used to predictably and continuously initiate fusion chain reactions that yield more energy than that incident on the fuel from the driver for the ultimate purpose of producing electrical power.

KD*P: Potassium dideuterium phosphate, a material widely used in frequency conversion optics.

Krypton fluoride (KrF) laser: Gas laser that operates in the ultraviolet at 248 nm.

Laser–plasma instability: Secondary processes such as symmetry disturbances, fuel preheat, and diversion of laser energy that occur when intense lasers interact with plasmas.

Liquid wall: Fusion reactor chamber's first wall that features thick jets of liquid coolant. This design may also shield the solid chamber walls from neutron damage.

Magnetized target fusion: ICF technique whereby a magnetic field is created surrounding the target; the field is then imploded around the target, initiating fusion reactions.

Mix (plasma physics): When colder target material is incorporated into the hot reaction region of the target, usually as a result of hydrodynamic instabilities.

Pulse compression: Technique whereby the incident pulse is compressed to deliver the energy in a shorter time.

Pulsed-power fusion: ICF technique that uses a large electrical current to magnetically implode a target.

Reactor chamber: Apparatus in which the fusion reactions would take place in an IFE power plant. It would contain and capture the energy released from repeated ignition.

Sabot: Protective device used when injecting an IFE target into the chamber at high speed.

Shock ignition: ICF technique that uses hydrodynamic shocks to ignite the compressed hot spot.

Target: Fuel capsule, together with a holhraum or other energy-focusing device (if one is used), that is struck by the driver's incident energy in order to initiate fusion reactions.

Wall-plug efficiency: Energy conversion efficiency defined as a ratio of the total driver output power to the input electrical power.

Wetted wall: Fusion reactor chamber's first wall that features a renewable, thin layer of liquid.

ACRONYMS

APG	advanced phosphate glass
AWE	Atomic Weapons Establishment
BOP	balance of plant
CEA	Commissariat à l'energie atomique
CELIA	Centre lasers intenses et applications
COE	cost of electricity
CPA	chirped-pulse amplification
CPP	continuous-phase plate
CVD	chemical vapor deposition
D	deuterium
DD (drive context)	direct drive
DEMO	demonstration plant
DOE	Department of Energy
DPSSL	diode-pumped solid-state laser
DT	deuterium-tritium
ELI	Extreme Light Infrastructure
ETF	engineering test facility
FAIR	Facility for Antiproton and Ion Research

FESAC	Fusion Energy Sciences Advisory Committee
FLiBe	fluorine-lithium-beryllium
FTF	Fusion Test Facility
GDP	glow discharge polymer
HAPL	High Average Power Laser
HCX	High-Current Experiment
HIF	heavy-ion fusion
HIFTF	Heavy-Ion Fusion Test Facility
HIF-VL	Heavy-Ion Fusion Virtual Laboratory
HI-IFE	Heavy-Ion Inertial Fusion Energy
HiPER	High-Power Laser Energy Research
HLW	high-level waste
ICF	inertial confinement fusion
ID	indirect drive
IFE	inertial fusion energy
i-LIFT	Laboratory Inertial Fusion Test
IRE	integrated research experiment
ISI	incoherent spatial imaging
ITER	International Thermonuclear Experimental Reactor
KDP	potassium dihydrogen phosphate
LANL	Los Alamos National Laboratory
LBNL	Lawrence Berkeley National Laboratory
LDRD	laboratory-directed research and development
LIFE	laser inertial fusion energy
LIL	Ligne d'Integration Laser
LLE	Laboratory for Laser Energetics
LLNL	Lawrence Livermore National Laboratory
LLW	low-level waste
LMJ	Laser MégaJoule (project)
LPI	laser–plasma interactions/instabilities
LTD	linear transformer driver
LULI	Laboratoire pour l'utilisation des lasers intenses
LWR	light water reactor
MagLIF	magnetized liner inertial fusion
MFE	magnetic fusion energy

MTF	Magnetized Target Fusion	
NCDX-II	neutralized drift compression experiment II	
NGNP	next-generation nuclear plant	
NIC	National Ignition Campaign	
NIF	National Ignition Facility	
NNSA	National Nuclear Security Administration	
NRC	National Research Council	
NRL	Naval Research Laboratory	
OFES	Office of Fusion Energy Sciences	
PALS	Prague Asterisk Laser System	
PDD	polar direct drive	
PETAL	PETawatt Aquitaine Laser	
PP	pulsed power	
PPPL	Princeton Plasma Physics Laboratory	
RF	radio frequency	
RTL	recyclable transmission line	
SAC	science advisory committee	
SAL	specific activity limit	
SBS	stimulated Brillouin scattering	
SNL	Sandia National Laboratories	
SRS	stimulated Raman scattering	
SSD	smoothing by spectral dispersion	
T	tritium	
TA	technology application	
TBM	test blanket module	
TPD	two-plasmon decay	
TRL	technology readiness level	
TWAC	TeraWatt ACcelerator	
UV	ultraviolet	
VLT	Virtual Laboratory for Technology	
YAG	yttrium-aluminum-garnet	

H

Summary from the Report of the Panel on the Assessment of Inertial Confinement Fusion (ICF) Targets (Unclassified Version)

The text below is excerpted from National Research Council, *Assessment of Inertial Confinement Fusion Targets* (The National Academies Press, Washington, D.C., 2013).

SUMMARY

In the fall of 2010, the Office of the U.S. Department of Energy's (DOE's) Under Secretary for Science asked for a National Research Council (NRC) committee to investigate the prospects for generating power using inertial fusion energy (IFE), noting that a key test of viability for this concept—ignition[1]—could be demonstrated at the National Ignition Facility (NIF) at Lawrence Livermore National Laboratory (LLNL) in the relatively near term. In response, the NRC formed both the Committee on the Assessment of the Prospects for Inertial Fusion Energy ("the committee") to investigate the overall prospects for IFE in an unclassified report and the separate Panel on Fusion Target Physics ("the panel") to focus on issues specific to fusion targets, including the results of relevant classified experiments and classified information on the implications of IFE targets for the proliferation of nuclear weapons.

This is the report of the Panel on Fusion Target Physics, which is intended to feed into the broader assessment of IFE being done by the NRC committee. It

---

[1] The operative definition of ignition adopted by the panel, "gain greater than unity," is the same as that used in the earlier NRC report *Review of the Department of Energy's Inertial Confinement Fusion Program*, Washington, D.C.: National Academy Press (1997).

consists of an unclassified body, which contains all of the panel's conclusions and recommendations, as well as three classified appendices, which provide additional support and documentation.

BACKGROUND

Fusion is the process by which energy is produced in the sun, and, on a more human scale, is the one of the key processes involved in the detonation of a thermonuclear bomb. If this process could be "tamed" to provide a controllable source of energy that can be converted to electricity—as nuclear fission has been in currently operating nuclear reactors—it is possible that nuclear fusion could provide a new method for producing low-carbon electricity to meet U.S. and world growing energy needs.

For inertial fusion to occur in a laboratory, fuel material (typically deuterium and tritium) must be confined for an adequate length of time at an appropriate density and temperature to overcome the Coulomb repulsion of the nuclei and allow them to fuse. In inertial confinement fusion (ICF)—the concept investigated in this report[2]—a driver (e.g., a laser, particle beam, or pulsed magnetic field) delivers energy to the fuel target, heating and compressing it to the conditions required for ignition. Most ICF concepts compress a small amount of fuel directly to thermonuclear burn conditions (a hot spot) and propagate the burn via alpha particle deposition through adjacent high-density fuel regions, thereby generating a significant energy output.

There are two major concepts for inertial confinement fusion target design: direct-drive targets, in which the driver energy strikes directly on the fuel capsule, and indirect-drive targets, in which the driver energy first strikes the inside surface of a hollow chamber (a hohlraum) surrounding the fuel capsule, producing energetic X-rays that compress the fuel capsule. Conventional direct and indirect drive share many key physics issues (e.g., energy coupling, the need for driver uniformity, and hydrodynamic instabilities); however, there are also issues that are unique to each concept.

The only facility in the world that was designed to conduct ICF experiments that address the ignition scale is the NIF at LLNL. The NIF driver is a solid-state laser. For the first ignition experiments, the NIF team has chosen indirect-drive targets. The NIF can also be configured for direct drive. In addition, important work on laser-driven, direct-drive targets (albeit at less than ignition scale) is also under way in the United States at the Naval Research Laboratory and the OMEGA

---

[2] Inertial confinement fusion (ICF) is the process by which the target is heated and compressed by the driver to reach fusion conditions. Inertial fusion energy (IFE) is the process by which useful energy is extracted from ignition and burn of ICF fuel targets.

laser at the University of Rochester. Heavy-ion-beam drivers are being investigated at the Lawrence Berkeley National Laboratory (LBNL), LLNL, and the Princeton Plasma Physics Laboratory (PPPL), and magnetic implosion techniques are being explored on the Z machine at Sandia National Laboratories (SNL) and at Los Alamos National Laboratory (LANL). Important ICF research is also under way in other countries, as discussed later in this report.

SPECIFIC CONCLUSIONS AND RECOMMENDATIONS

The panel's key conclusions and recommendations, all of them specific to various aspects of inertial confinement fusion, are presented below. They are labeled according to the chapter and number order in which they appear in the text, to provide the reader with an indicator of where to find a more complete discussion. This summary ends with two overarching conclusions and an overarching recommendation derived from viewing all of the information presented to the panel as a whole.

Targets for Indirect Laser Drive

CONCLUSION 4-1: The national program to achieve ignition using indirect laser drive has several physics issues that must be resolved if it is to achieve ignition. At the time of this writing, the capsule/hohlraum performance in the experimental program, which is carried out at the NIF, has not achieved the compressions and neutron yields expected based on computer simulations. At present, these disparities are not well understood. While a number of hypotheses concerning the origins of the disparities have been put forth, it is apparent to the panel that the treatments of the detrimental effects of laser-plasma interactions (LPI) in the target performance predictions are poorly validated and may be very inadequate. A much better understanding of LPI will be required of the ICF community.

CONCLUSION 4-2: Based on its analysis of the gaps in current understanding of target physics and the remaining disparities between simulations and experimental results, the panel assesses that ignition using laser indirect drive is not likely in the next several years.

The National Ignition Campaign (NIC) plan—as the panel understands it—suggests that ignition is planned after the completion of a tuning program lasting 1-2 years that is presently under way and scheduled to conclude at the end of FY2012. While this success-oriented schedule remains possible, resolving the present issues and addressing any new challenges that might arise are likely to push the timetable for ignition to 2013-2014 or beyond.

Targets for Indirect-Drive Laser Inertial Fusion Energy

CONCLUSION 4-4: The target design for a proposed indirect-drive IFE system (the Laser Inertial Fusion Energy, or LIFE, program developed by LLNL) incorporates plausible solutions to many technical problems, but the panel assesses that the robustness of the physics design for the LIFE target concept is low.

- The proposed LIFE target presented to the panel has several modifications relative to the target currently used in the NIC (e.g., rugby hohlraums, shine shields, and high-density carbon ablators), and the effects of these modifications may not be trivial. For this reason, R&D and validation steps would still be needed.
- There is no evidence to indicate that the margin in the calculated target gain ensures either its ignition or sufficient gain for the LIFE target. If ignition is assumed, then the gain margin briefed to the panel, which ranged from 25 percent to almost 60 percent when based on a calculation that used hohlraum and fuel materials characteristic of the NIC rather than the LIFE target, is unlikely to compensate for the phenomena relegated to it—for example, the effects of mix—under any but the most extremely favorable eventuality. In addition, the tight coupling of LIFE to what can be tested on the NIF constrains the potential design space for laser-driven, indirect-drive IFE.

Targets for Direct-Drive Laser Inertial Fusion Energy

CONCLUSION 4-6: The prospects for ignition using laser direct drive have improved enough that it is now a plausible alternative to laser indirect drive for achieving ignition and for generating energy.

- The major concern with laser direct drive has been the difficulty of achieving the symmetry required to drive such targets. Advances in beam-smoothing and pulse-shaping appear to have lessened the risks of asymmetries. This assessment is supported by data from capsule implosions (performed at the University of Rochester's OMEGA laser), but it is limited by the relatively low drive energy of the implosion experiments that have thus far been possible. Because of this, the panel's assessment of laser-driven, direct-drive targets is not qualitatively equivalent to that of laser-driven, indirect-drive targets.
- Further evaluation of the potential of laser direct-drive targets for IFE will require experiments at drive energies much closer to the ignition scale.
- Capsule implosions on OMEGA have established an initial scaling point that indicates the potential of direct-drive laser targets for ignition and high yield.

- Polar direct-drive targets[3] will require testing on the NIF.
- Demonstration of polar-drive ignition on the NIF will be an important step toward an IFE program.
- If a program existed to reconfigure the NIF for polar drive, direct-drive experiments that address the ignition scale could be performed as early as 2017.

Fast Ignition

Fast ignition (FI) requires a combination of long-pulse (implosion) and short-pulse (ignition) lasers. Aspects of fast ignition by both electrons and protons were briefed to the panel. Continued fundamental research into fast ignition theory and experiments, the acceleration of electrons and ions by ultrashort-pulse lasers, and related high-intensity laser science is justified. However, issues surrounding low laser-target energy coupling, a complicated target design, and the existence of more promising concepts (such as shock ignition) led the panel to the next conclusion regarding the relative priority of fast ignition for fusion energy.

CONCLUSION 4-5: At this time, fast ignition appears to be a less promising approach for IFE than other ignition concepts.

Laser-Plasma Interactions

A variety of LPI take place when an intense laser pulse hits the target capsule or surrounding hohlraum. Undesirable effects include backscattering of laser light, which can result in loss of energy; cross-beam energy transfer among intersecting laser beams, which can cause loss of energy or affect implosion symmetry; acceleration of suprathermal "hot electrons," which then can penetrate and preheat the capsule's interior and limit later implosion; and filamentation, a self-focusing instability that can exacerbate other LPI. LPI have been a key limiting factor in laser inertial confinement fusion, including the NIC indirect-drive targets, and are still incompletely understood.

CONCLUSION 4-11: The lack of understanding surrounding laser-plasma interactions remains a substantial but as yet unquantified consideration in ICF and IFE target design.

---

[3] In polar direct drive, the driver beams are clustered in one or two rings at opposing poles. To increase the uniformity of the drive, polar drive beams strike the capsule obliquely, and the driver energy is biased in favor of the more equatorial beams.

RECOMMENDATION 4-1: DOE should foster collaboration among different research groups on the modeling and simulation of laser-plasma interactions.

Heavy-Ion Targets

A wide variety of heavy-ion target designs has been investigated, including indirect-drive, hohlraum/capsule targets that resemble NIC targets. Recently, the emphasis has shifted to direct-drive targets, but to date the analysis of how these targets perform has been based on computation rather than experiment, and the codes have not been benchmarked with experiments in relevant regimes.

CONCLUSION 4-12: The U.S. heavy-ion-driven fusion program is considering direct-drive and indirect-drive target concepts. There is also significant current work on advanced target designs.[4] This work is at a very early stage, but if successful may provide very high gain.

- The work in the heavy-ion fusion (HIF) program involves solid and promising science.
- Work on heavy-ion drivers is complementary to the laser approaches to IFE and offers a long-term driver option for beam-driven targets.
- The HIF program relating to advanced target designs is in a very early stage and is unlikely to be ready for technical assessment in the near term.
- The development of driver technology will take several years, and the cost to build a significant accelerator driver facility for any target is likely to be very high.

Z-Pinch Targets

Current Z-pinch direct-drive concepts utilize the pressure of a pulsed, high magnetic field to implode deuterium-tritium fuel to fusion conditions. Simulations predict that directly using the pressure of the magnetic field to implode and compress the target can greatly increase the efficiency with which the electrical energy is coupled to the fuel as compared with the efficiency of indirect drive from Z-pinch X-ray sources. There is work under way on both classified and unclassified target designs.

CONCLUSION 4-13: Sandia National Laboratories is leading a research effort on a Z-pinch scheme that has the potential to produce high gain with good energy

---

[4] Advanced designs include direct-drive, conical X-target configurations (see Chapter 2).

efficiency, but concepts for an energy delivery system based on this driver are too immature to be evaluated at this time.

It is not yet clear that the work at SNL will ultimately result in the high gain predicted by computer simulations, but initial results are promising and it is the panel's opinion that significant progress in the physics may be made in a year's time. The pulsed-power approach is unique in that its goal is to deliver a large amount of energy (~10 MJ) to targets with good efficiency (≥10 percent) and to generate large fusion yields at low repetition rates.

Target Fabrication

Current targets for inertial confinement fusion experiments tend to be one-off designs, with specifications that change according to the experiments being run. In contrast, targets for future IFE power plants will have to have standard, low-cost designs that are mass-produced in numbers as high as a million targets per day per power plant. The panel examined the technical feasibility of producing targets for various drivers, including limited aspects of fabrication for IFE. However, a full examination of the issues of mass production and low cost is the province of the NRC IFE committee study.

CONCLUSION 4-7: In general, the science and engineering of manufacturing fusion targets for laser-based ICF are well advanced and meet the needs of those experiments, although additional technologies may be needed for IFE. Extrapolating this status to predict the success of manufacturing IFE targets is reasonable if the target is only slightly larger than the ICF target and the process is scalable. However, subtle additions to the design of the ICF target to improve its performance (greater yield) and survivability in an IFE power plant may significantly affect the manufacturing paradigm.

Proliferation Risks of IFE

Many modern nuclear weapons rely on a fusion stage as well as a fission stage, and there has been discussion of the potential for host state proliferation—particularly vertical proliferation—associated with the siting of an IFE power plant. The panel was asked to evaluate the proliferation risks associated with IFE, particularly with regard to IFE targets.

CONCLUSION 3-1: At present, there are more proliferation concerns associated with indirect-drive targets than with direct-drive targets. However, the spread of technology around the world may eventually render these concerns moot. Remaining concerns are likely to focus on the use of classified codes for target design.

CONCLUSION 3-2: The nuclear weapons proliferation risks associated with fusion power plants are real but are likely to be controllable. These risks fall into three categories:

- **Knowledge transfer,**
- **Special nuclear material (SNM) production, and**
- **Tritium diversion.**

OVERARCHING CONCLUSIONS AND RECOMMENDATION

While the focus of this panel was on ICF target physics, the need to evaluate driver-target interactions required considering driver characteristics as well. This broader analysis led the panel to the following overarching conclusions and a recommendation.

OVERARCHING CONCLUSION 1: The NIF has the potential to support the development and further validation of physics and engineering models relevant to several IFE concepts, from indirect-drive hohlraum designs to polar direct-drive ICF and shock ignition.

- **In the near to intermediate term, the NIF is the only platform that can provide information relevant to a wide range of IFE concepts at ignition scale. Insofar as target physics is concerned, it is a modest step from NIF scale to IFE scale.**
- **Targets for all laser-driven IFE concepts (both direct-drive and indirect-drive) can be tested on the NIF. In particular, reliable target performance would need to be demonstrated before investments could confidently be made in the development of laser-driven IFE target designs.**

The NIF will also be helpful in evaluating indirectly driven, heavy-ion targets. It will be less helpful in gathering information relevant to current Z-pinch, heavy-ion direct drive, and heavy-ion advanced target concepts.

OVERARCHING CONCLUSION 2: It would be advantageous to continue research on a range of IFE concepts, for two reasons:

- **The challenges involved in the current laser indirect-drive approach in the single-pulse National Nuclear Security Administration program at the NIF have not yet been resolved, and**
- **The alternatives to laser indirect drive have technical promise to produce high gain.**

In particular, the panel concludes that laser direct drive is a viable concept to be pursued on the NIF. SNL's work on Z-pinch can serve to mitigate risk should the NIF not operate as expected. This work is at a very early stage but is highly complementary to the NIF approach, because none of the work being done at SNL relies on successful ignition at the NIF, and key aspects of the target physics can be investigated on the existing Z-machine. Finally, emerging heavy-ion designs could be fruitful in the long term.

OVERARCHING RECOMMENDATION: The panel recommends against pursuing a down-select decision for IFE at this time, either for a specific concept such as LIFE or for a specific target type/driver combination.

Further R&D will be needed on indirect drive and other ICF concepts, even following successful ignition at the NIF, to determine the best path for IFE in the coming decades.

I

Technical Discussion of the Recent Results from the National Ignition Facility

The Lawson criterion for ignition[1,2] requires that the product $P\tau$ exceeds a threshold value that depends on the plasma temperature. The central temperature of an ICF imploded capsule is roughly proportional to the capsule's implosion velocity. The implosion velocity is limited to values below ~400 km/s to prevent hydrodynamic instabilities from breaking up the imploding shell. This constraint on the implosion velocity keeps the central temperature at ~5 keV. At such relatively low temperature, the onset of ignition requires[3] a product $P\tau$ exceeding ~30 Gbar-ns. Using the results of Betti et al.[4] applied to NIC experiments, current implosions have achieved $P\tau$ ~ 10-18 Gbar-ns[5] and a temperature of 3-4 keV. The highest $P\tau$, ~18 Gbar-ns, is about half of the ignition requirement. Time-resolved measurements of the compressed core X-ray emission indicate that the confinement time τ is 100-150 ps, suggesting that pressures of 100-130 Gbar have been achieved.[6] To achieve ignition-relevant $P\tau \geq 30$ Gbar-ns, pressures exceeding 300 Gbar are required.

---

[1] J.D. Lawson, 1957, *Proceedings of the Physical Society London B* 70: 6.
[2] R. Betti, P.Y. Chang, B.K. Spears, et al., 2010, Thermonuclear ignition in inertial confinement fusion and comparison with magnetic confinement, *Physics of Plasmas* 17: 058102.
[3] Ibid.
[4] Ibid.
[5] S. Glenzer, D.A. Callahan, A.J. MacKinnon, et al., 2012, Cryogenic thermonuclear fuel implosions on the National Ignition Facility, *Physics of Plasmas* 19: 056318, and R. Betti, 2012, "Theory of Ignition and Hydroequivalence for Inertial Confinement Fusion, Overview Presentation," OV5-3, 24th IAEA Fusion Energy Conference, October 7-12, San Diego, Calif.
[6] Ibid.

The compressed core of an ICF implosion consists of a central hot plasma (the hot spot) surrounded by a cold dense shell. The total areal density determines the hot spot confinement by the surrounding dense shell. The NIF indirect-drive point design target is intended to implode at low entropy to produce high areal densities. To date, the highest areal density measured in the experiments was 1.25 g/cm^2 (shot N120321), about 20 percent below the design value of 1.5 g/cm^2. The areal density of the central hot spot is another important parameter because it determines the capacity of the hot spot to slow down the 3.5-MeV fusion alpha particles required to trigger the ignition process. Hot spot areal densities up to ~70 mg/cm^2 have been inferred from the measurements of the neutron yields, hot spot size, ion temperature, and burn duration. Such values of the hot spot areal densities are enough to slow down more than 50 percent of the alpha particles at the low temperatures (~3-4 keV) measured in the experiments but are not sufficient for ignition since alpha particles need to be slowed down at higher temperatures between 5 and 10 keV. At these high temperatures, the hot spot areal density needs to exceed ~200 mg/cm^2 to stop the fusion alphas. The highest temperature achieved to date is ~4 keV, which is close to the ~5 keV required for the onset of ignition. However, in the experiments, the highest temperature and highest areal densities were not achieved on the same implosion. The temperature was ~3 keV in the highest areal density implosion to date.

Together with the areal density, pressure, and temperature, the neutron yield is a critical parameter determining the performance of an implosion. A rough estimate of the expected neutron yield from the compression alone, without accounting for alpha particle heating, in the absence of nonuniformities—that is, a one-dimensional (1-D), or clean, implosion—can be obtained from a simple formula[7] relating the yield to the measured areal density and ion temperature by $Y_n^{16} \approx \rho R^{0.56} (T/4.7)^{4.7} M_{DT} / 0.24$, where the neutron yield, Y_n^{16}, is expressed in units of 10^{16}, the areal density ρR is in g/cm^2, the temperature T in keV, and the DT mass M_{DT} in mg.

A straightforward substitution of $\rho R = 1$ g/cm^2, T = 4 keV, and $M_{DT} = 0.17$ mg leads to a compression 1-D yield of 3.3×10^{15} neutrons, about 4-8 times higher than currently measured in the experiments ($4 - 9 \times 10^{14}$).

An overall performance parameter used by the LLNL group is the experimental ignition threshold factor (ITFx).[8] The ITFx has been derived by fitting the results of hundreds of computer simulations of ignition targets to find a measurable

---

[7] R. Betti, P.Y. Chang, B.K. Spears, et al., 2010, Thermonuclear ignition in inertial confinement fusion and comparison with magnetic confinement, *Physics of Plasmas* 17: 058102.

[8] B. Spears, S. Glenzer, M.J. Edwards, et al., 2012, Performance metrics for inertial confinement fusion implosions: Aspects of the technical framework for measuring progress in the National Ignition Campaign, *Physics of Plasmas* 19: 056316.

parameter indicative of the performance with respect to ignition. An implosion with ITFx = 1 has a 50 percent probability of ignition. It can be shown[9] that the ITFx represents the third power of the Lawson criterion ITFx = $[(P\tau)/(P\tau)_{ig}]^3$, where $(P\tau)_{ig}(T)$ is a function of temperature, representing the minimum product $P\tau$ required for ignition at a given temperature.[10] For the indirect-drive point-design target with 0.17 mg of DT fuel, the ITFx can be expressed[11] in terms of the measured areal density and neutron yield according to

$$\text{ITFx} \approx \left(\frac{\rho R}{1.5}\right)^{2.3}\left(\frac{Y_n^{16}}{0.32}\right)$$

Both the areal density and neutron yield are the so-called no-burn or no-alpha values as they are related solely to the hydrodynamic compression without accounting for alpha particle energy deposition. To date, the highest value of the ITFx is about 0.1 from implosions, with areal densities and neutron yields in the range 0.8-1.2 g/cm² and 5-8 × 10¹⁴ respectively.[12]

---

[9] R. Betti, 2012, "Theory of Ignition and Hydroequivalence for Inertial Confinement Fusion, Overview Presentation," OV5-3, 24th IAEA Fusion Energy Conference, October 7-12, San Diego, Calif.; B. Spears, S. Glenzer, M.J. Edwards, et al., 2012, Performance metrics for inertial confinement fusion implosions: Aspects of the technical framework for measuring progress in the National Ignition Campaign, *Physics of Plasmas* 19: 056316.

[10] R. Betti, P.Y. Chang, B.K. Spears, et al., 2010, Thermonuclear ignition in inertial confinement fusion and comparison with magnetic confinement, *Physics of Plasmas* 17: 058102; R. Betti, "Theory of Ignition and Hydroequivalence for Inertial Confinement Fusion, Overview Presentation," OV5-3, 24th IAEA Fusion Energy Conference, October 7-12, San Diego, Calif.

[11] B. Spears, S. Glenzer, M.J. Edwards, et al., 2012, Performance metrics for inertial confinement fusion implosions: Aspects of the technical framework for measuring progress in the National Ignition Campaign, *Physics of Plasmas* 19: 056316.

[12] S. Glenzer, D.A. Callahan, A.J. MacKinnon, et al., 2012, Cryogenic thermonuclear fuel implosions on the National Ignition Facility, *Physics of Plasmas* 19: 056318; J. Edwards, et al., 2012, "Progress Towards Ignition on the National Ignition Facility," MR1.00001, 54th Annual Meeting of the American Physical Society, Division of Plasma Physics, Philadelphia, Pa., October 29-November 2.

J

Detailed Discussion of Technology Applications Event Profiles

The following narratives will indicate the steps required for each technology application (TA) to reach the starting point of the DEMO conceptual design. Conceptual design of DEMO reactors will depend on one or more TAs successfully achieving technology readiness levels (TRLs) of 6 for each component of that TA "package." The specific steps are meant to be illustrative of the conditional requirements that DOE should set down in its planning process—requirements that should be regularly updated based on scientific and technological progress.

LASER IFE EVENTS-BASED ROADMAP TO DEMO (TA-1)

In addition to the target gain and laser efficiency demonstrations required before operation of an FTF or design of a DEMO reactor, additional detailed pre-conditions are required for each of three main laser IFE candidate technology applications (TAs).

Indirect Drive Target with Diode-Pumped Laser: Pre-conditions for FTF or DEMO

1a. In the present National Ignition Facility (NIF) indirect drive campaign, if $1 < G < 10$ is achieved, there should be a further program of work on NIF to extend the gain well into the reactor-scale range before committing to an FTF or DEMO.

1b. If G < 1 is the final result of the National Ignition Campaign (NIC) and follow-on campaigns after some reasonable period of scientific testing, then other drive approaches should be investigated as planned.

1c. The diode-pumped solid-state laser is optically very similar to the flashlamp-pumped NIF laser and so experiments on NIF will define future expectations for indirect drive with a diode-pumped laser. Assuming G > 10, before commitment to an FTF or DEMO, the following achievements will be necessary simultaneously in one laser IRE device, for instance,

—Energy in the 5 kJ range in the ultraviolet as planned.
—Efficiency >10 percent with 15 percent goal in UV.
—Repetition frequency >5 Hz, with clear technical extension to >15 Hz.
—Life test to >10^7 pulses, with clear technical extension to >10^9 pulses using the same medium.

1d. A chamber design with life expectancy of >10^8 pulses must exist for the indirect-drive threat spectrum and the chamber design to include final optical elements.

1e. Target fabrication must project to the precision and economy required of reactor operation.

Direct-Drive Target with Diode-Pumped Laser: Pre-conditions for FTF or DEMO

As with indirect drive, the diode-pumped laser will be optically very similar to the flashlamp-pumped NIF laser, and so laser performance on the NIF will define future expectations for direct drive with a diode-pumped laser.

Regardless of the outcome on indirect drive, even in the case that reactor-scale gain is achieved (1a above), the NIF laser should be used to study direct-drive targets, as planned.

Polar direct drive (PDD) is an interim approach to spherical direct drive (SDD) that employs the existing NIF beam ports. However, ignition with PDD is uncertain owing to likely laser-plasma instability (LPI) differences between the "equatorial" and the more polar beams. Polar direct drive may be a valid test bed for a preview of spherical direct-drive interactions on the NIF laser.

2a. In event 1b above, with G < 1 in indirect drive at the end of the ignition campaign, NIF should be upgraded as planned for PDD studies (2017) with beam smoothing (estimated $30 million for materials) and employed in a study of PDD physics at reactor plasma-scale size. If modeling of the results with validated codes points to likely G > 1 with SDD, the NIF should be reconfigured at the earliest opportunity to a true SDD configuration (estimated $300 million).

2b. If 1 < G < 10 is achieved with SDD on the NIF there should be additional work to tune as far as possible to reactor-scale gains.

2c. Until the SDD and ID approaches on the NIF both fail to achieve 1 < G < 10 in item 2b, the diode-pumped solid-state laser should continue to be developed. Before commitment to an FTF or DEMO, assuming G > 10 is achieved, all of the following achievements are needed simultaneously in one DPSSL laser IFE beam line:

—Energy in the 5 kJ range in the ultraviolet as planned.
—Efficiency >10 percent with 15 percent goal in the UV, as planned.
—Repetition frequency >5 Hz, with clear technical extension to >15 Hz.
—Life test to >10^7 pulse, with clear technical extension to >10^9 pulses using the same medium.

2d. A chamber design with life expectancy >10^8 pulses must exist for the direct-drive threat spectrum and the chamber design must include final optical elements.

2e. Target fabrication must project to the precision and economy required of reactor operation.

Direct-Drive Target with KrF Laser: Preconditions for FTF or DEMO

There is not an ignition-level facility available at the KrF wavelength of 248 nm with bandwidth of 3 THz. However, calculations presented to the committee based on spherical direct drive predict the lowest energy threshold for ignition to occur with KrF. These calculations are plausible because the LPI threshold of KrF is higher by a factor of 2 compared to 3ω thresholds at 351 nm. This potential benefit of KrF suggests that, if reactor-scale gain of 140 is achieved under heading 2b above, cost-effective power generation could be possible with KrF-driven IFE.

Prior to construction and operation of a 400-500 kJ KrF laser FTF for the exploration of SDD physics with reactor-scale targets at 248 nm, the committee suggests the following preconditions to maximize the chance that power generation by KrF-driven, direct-drive IFE will be cost competitive.

3a. A single-shot 15-25 kJ KrF beamline operates at 0.01 Hz with the desired pulse shape, focal uniformity, and zooming (~20 copies of this beamline would drive the facility).

3b. The NRL Electra repetitive test of a 500 J KrF laser at 5 Hz runs for >10^7 pulses with efficiency of >6 percent and a clear projection of the same technology to the 15-25 kJ module at >10^9 pulses.

3c. Experimental evidence validates some aspects of high gain (>140) in 2D(+) calculations that include the most advanced validated models of LPI at 248 nm and incorporate learning from SDD experiments on the NIF.
3d. A chamber design exists that projects to >10^8 pulses with the threat spectrum of direct-drive targets, to include a plausible final optics design, and that direct drive targets can be injected into the chamber and engaged by the laser at a >5 Hz rate.
3e. Target manufacture projects to mass production at the quality desired for direct drive and within the cost required for power production.
3f. KrF direct-drive laser IFE is estimated to be cost-competitive with other IFE or MFE plant designs.

Note that the NIF can also be upgraded to operate at 4ω in the deep UV if such operation is necessary for testing LPI at the deep UV vs 351 nm.

HEAVY-ION IFE EVENTS-BASED ROADMAP TO DEMO (TA-2)

There are several technical approaches to heavy-ion inertial fusion. Each approach uses a particular kind of accelerator, a particular kind of target, and a particular kind of chamber. The two principal types of accelerators are radio-frequency (RF) accelerators and induction linear accelerators (linacs). Unlike laser fusion, there is nearly a continuum of targets ranging from targets that are fully directly driven to targets that are indirectly driven. Ultimately, the program must determine the optimal point in this continuum, but, in this section, we will simply distinguish between direct drive and indirect drive. As is the case for lasers, the target ignition modes include hot-spot ignition, shock ignition, and fast ignition. Heavy-ion fusion appears to be compatible with several types of chambers, but most power plant studies have adopted chambers with thick liquid walls to minimize radiation damage to materials.

In order to make progress on limited funds there has, for many years, been an informal agreement that the United States would pursue induction linacs while the foreign programs would pursue RF accelerators. In the near term it is not necessary to choose between direct drive and indirect drive. The accelerator requirements for the two cases are similar. The accelerator requirements for fast ignition are quite different. Fast ignition targets require high kinetic energy ions compared to other types of targets. The large RF heavy-ion accelerators in Germany and Russia are designed to produce high kinetic energies. Fast ignition is an important part of some of these foreign programs. Although large future machines such as the Facility for Antiproton and Ion Research (FAIR) in Germany may be able to do some preliminary experiments on fast ignition, they will likely fall short of the required ignition temperature by more than two orders of magnitude. Consequently it appears difficult to validate ion fast ignition physics. In the remainder of this section the committee considers only the U.S. program—induction linacs and direct or indirect drive.

Pre-conditions for FTF or DEMO

Much of the target information for heavy-ion fusion is based on computer simulations using the codes that are also used for laser and pulsed power fusion. There is also limited experimental information on ion-driven fusion, including heavy-ion energy deposition experiments in cold and laser-heated matter and light-ion-beam-driven hohlraum data up to about 60 eV.[1,2] For information on inertial confinement fusion physics, it is currently necessary to rely on classified data and the laser fusion programs, particularly the NIF program. Given this situation, the committee now turns to the pre-conditions needed for a heavy-ion fusion FTF or DEMO:

1a. Laboratory-scale ignition on NIF or elsewhere is necessary. These ignition experiments must be convincingly connected, using state-of-the-art computer simulations and existing ion target data, to the achievement of high gain (G > 30) ion-driven targets. Since the fuel capsules for indirectly driven ion-beam fusion are similar or identical to those for indirectly driven laser fusion, and since ions have driven hohlraums to approximately 60 eV, it is much easier to make a convincing connection for indirect drive than for direct drive.

1b. In addition to the current uncertainties in target physics, there are also uncertainties in accelerator physics, at least for the high current beams needed for fusion. To address these uncertainties it is necessary to show that NDCX-II, the ion induction linac currently coming on line at the Lawrence Berkeley National Laboratory (LBNL), meets its designs goals and that its performance matches theory and simulation. A result of these experiments should be a validation of the accelerator and beam physics codes at increasing intensity.

1c. Transport of driver-scale beam charge density in magnetic quadrupoles without serious degradation of beam quality (ability to be focused) must be demonstrated and provide further validation for beam transport codes. This can be done by restarting and upgrading the existing HCX accelerator at LBNL.

1d. Ion sources, magnetic quadrupole arrays, high-gradient insulators, high-voltage pulsers (similar to those needed for the KrF and pulsed power approaches

---

[1] T.A. Mehlhorn, 1997, Intense ion beams for inertial confinement fusion, *IEEE Transactions on Plasma Science* 25(6): 1336-1356.

[2] M.S. Derzon, G.A. Chandler, R.J. Dukart, D.J. Johnson, R.J. Leeper, M.K. Matzen, E.J. McGuire, T.A. Mehlhorn, A.R. Moats, R.E. Olson, and C.L. Ruiz, 1996, Li-beam-heated hohlraum experiments at particle beam fusion accelerator II, *Physical Review Letters* 76: 435-438.

to IFE), and magnetic materials for induction cores must be further developed to demonstrate adequate cost, reliability, durability, voltage gradient, and efficiency. These components must be assembled into induction acceleration units in an IRE. Pulsing these units at 10 Hz for 3 years will give a total of approximately 10^9 shots of reliability and durability testing.

1e. It is necessary to produce a complete design of a final focusing system that rigorously meets all known requirements associated with beam physics and shielding. This focusing system must be integrated with a credible chamber design.

1f. The successful completion of items 1a through 1e leads to a major decision point, the decision to proceed with the construction of a 10-kJ to 100-kJ accelerator, the initial step of an FTF. This accelerator must validate the performance of scaled hohlraums and/or adequate hydrodynamic stability for directly driven ion targets. If the estimated cost of this facility is greater than a few hundred million dollars, item 1d has failed to demonstrate adequate cost since the cost of this facility would not extrapolate to acceptable cost for a full-scale driver.

1g. If the intermediate accelerator described in 1f successfully validates the target physics for direct and/or indirect drive and if credible target fabrication techniques and a credible chamber have been successfully demonstrated, there is enough information to make a decision to construct a full-scale accelerator driver. This driver must demonstrate an efficiency-gain product ≥ 10. At this point, enough information would be available to proceed to an FTF. To minimize the cost of performing the demonstration of efficiency and gain, the driver would be built initially without all the power supplies necessary for high repetition rate. It would be upgraded to drive an FTF by adding more power supplies.

PULSED POWER IFE EVENTS-BASED ROADMAP TO DEMO (TA-3)

There are two technology applications (TAs) to pulsed power (PP) IFE at present. One involves magnetic implosion of magnetized, laser-preheated fusion fuel on a ~100 ns timescale and goes by the name of Magnetized Liner Inertial Fusion, or MagLIF. Other unpublished approaches that would use ~100 ns pulsed power to implode fusion fuel are also under consideration. The other TA, called Magnetized Target Fusion, or MTF, is related to MagLIF through the use of PP technology and magnetic implosion as the driver approach but is otherwise quite distinct: The implosion timescale is more than 10 times longer, the length scale is more than 10 times larger, the magnetic configuration is different (MTF seeks to compress a field-reversed configuration because of the longer timescale) and the plasma density is 100-1,000 times lower. In a broad IFE program including PP IFE, there would be one down-select based on physics and technology between the shorter and longer pulse PP IFE TAs.

Although the power-plant ideas presented by the proponents of MagLIF and MTF differ, the challenges are the same: high yield per pulse in a liquid wall chamber at a repetition rate of order 0.1 Hz; the chamber must be commercially viable and long-lived; and delivery of the current to the target must be accomplished reliably with standoff. Generically, the latter challenge is addressed with recyclable transmission lines (RTLs), and the chamber is assumed to be a thick liquid wall chamber that must recover "completely" to its undisturbed state in the ~10 s between pulses.

MagLIF Approach: Pre-conditions for FTF or DEMO

Up to now, all "data" on MagLIF is from computer simulations. A substantial systematic experimental campaign is planned each year for 5 years to validate the computer simulations and to determine if the goal of scientific breakeven can be achieved on the existing 27 MA Z-machine at Sandia. Scientific breakeven is defined as fusion energy out (using D-T fuel) equals energy delivered to the fuel.

1a. If scientific breakeven is achieved and predictive validity of the design code(s) is demonstrated, results should be compared with other existing results. If one is clearly making more progress than the other, a down-select might be made by the end of the 5-year period based on code predictions of which will be the most favorable approach for IFE. Here the committee assumes that it is unnecessary to take into account differences in reactor technology to do this down-selection. However, if there are significant differences, the necessary engineering design tasks should be carried out during the 5-year period. The conceptual design of a gain >1 facility should be developed. If possible, that facility should be designed to be upgradeable to a high gain facility (FTF) rather than requiring a completely new facility.

1b. If scientific breakeven is achieved but predictive capability is not achieved, experiments and theoretical research must continue before any decision is made to go for an IFE ignition facility. However, the National Nuclear Security Administration (NNSA) may decide to initiate preparations for a single-shot-ignition and high-gain facility depending on mission requirements.

1c. If scientific breakeven is not achieved and the reasons are not understood, MagLIF's place in the broad IFE program should be reconsidered in light of progress on other TAs.

1d. PP technology must have favorable long lifetime and high efficiency projections as well as low maintenance and repair cost expectations for MagLIF to proceed to an FTF, although a single-shot high-gain facility might still be of interest to NNSA.

1e. A conceptual chamber design with life expectancy of >10^7 pulses must exist for the 0.1 Hz, 10 GJ yields presently favored by PP IFE proponents, or the approach must be reoptimized at a different rep-rate and yield per pulse. Additionally, engineering projections for use of RTLs must be favorable and proof-of-principle experiments for their use in a PP system must be successful before an FTF design is undertaken.

MTF approach to PP IFE: Preconditions for FTF or DEMO.

Laboratory experiments on the Shiva Star (operating at 4.5 MJ) capacitor bank deliver up to 12 MA of current to a 10 cm diameter, 30 cm long, 1 mm thick aluminum (Al) cylinder. Assuming success of integrated experiments in which field reversed configuration plasmas are injected into the Al cylinder and then imploded, explosively driven experiments are to follow. Computer simulations are carried out using the Mach2 MHD code.

2a. The Shiva Star experiments are expected to achieve $>10^{19}/cm^3$, 3-5 keV, ~1 cm diameter plasmas confined in a 300-500 T (peak field) field-reversed plasma configuration in ~3 years. Success here would lead to the explosively driven implosion experiments, which could achieve breakeven. The success of the explosively driven experiments together with demonstrated predictive capability would make MTF a competitor at the time of PP IFE down-selection in about 5 years. "Predictive capability" means that the enhancement of yield due to the presence of magnetic field in the initial plasma should be understood in detail in spite of poor diagnostic access.

2b. If scientific breakeven is achieved but predictive capability is not achieved, experiments and theoretical research must continue before any decision is made to go for an IFE ignition facility.

2c. If scientific breakeven is not achieved and the reasons are not understood, MTF's place in the broad IFE program should be reconsidered in light of progress on other TAs.

2d. PP technology must have favorable long life-time and high efficiency projections as well as low maintenance and repair cost expectations in order for MTF to go on to an FTF, although a single-shot high-gain facility might still be of interest to NNSA.

2e. A conceptual chamber design with life expectancy of $>10^7$ pulses must exist for the 0.1 Hz, 5 GJ yields presently favored by MTF proponents. Additionally, engineering projections for use of RTLs must be favorable, and proof-of-principle experiments for their use in a PP system must be successful before an FTF design is undertaken.